THE SCIENCE OF

The
SCIENCE
of WAR

DEFENSE BUDGETING, MILITARY TECHNOLOGY,
LOGISTICS, AND COMBAT OUTCOMES

Michael E. O'Hanlon

Princeton University Press
Princeton and Oxford

IN ASSOCIATION WITH THE BROOKINGS INSTITUTION

Copyright © 2009 by Princeton University Press

Published by Princeton University Press, 41 William Street,
Princeton, New Jersey 08540
In the United Kingdom: Princeton University Press, 6 Oxford Street,
Woodstock, Oxfordshire OX20 1TW

press.princeton.edu

All Rights Reserved

Second printing, and first paperback printing, 2013
Paperback ISBN 978-0-691-15799-3

The Library of Congress has cataloged the cloth edition of this book as follows

O'Hanlon, Michael E.
The science of war : defense budgeting, military technology, logistics,
and combat outcomes / Michael E. O'Hanlon.
p. cm.
Includes index.
ISBN 978-0-691-13702-5 (hardcover)
1. Military planning—United States. 2. United States—
Defenses—Decision making. I. Title.
U153.O337 2009
355'.033573—dc22   2009001378

British Library Cataloging-in-Publication Data is available

This book has been composed in Adobe Caslon

Printed on acid-free paper. ∞

Printed in the United States of America

3 5 7 9 10 8 6 4 2

CONTENTS

Acknowledgments vii

INTRODUCTION 1

CHAPTER I
Defense Budgeting
and Resource Allocation 8

CHAPTER II
Modeling Combat and
Sizing Forces 63

CHAPTER III
Logistics and Overseas Bases 141

CHAPTER IV
Technical Issues in
Defense Analysis 169

CONCLUSION 243

APPENDIX
Figures and Tables 249

Index 263

ACKNOWLEDGMENTS

I WOULD LIKE TO THANK a number of friends, colleagues, professors, role models, and others who have helped me in one way or another, directly or indirectly, with this book. They include Adriana Lins de Albuquerque, Michael Berger, Richard Betts, Steve Biddle, Marty Binkin, Bruce Blair, Dennis Blair, Barry Blechman, Max Boot, Christopher Bowie, Duncan Brown, Richard Bush, Jason Campbell, Kurt Campbell, Ash Carter, Eliot Cohen, William Cohen, Bernard Cole, Walter Cronkite, Ivo Daalder, Thomas Davis, Todd Duso, Joshua Epstein, Harold Feiveson, Michele Flournoy, Lawrence Freedman, Aaron Friedberg, Richard Garwin, Thomas Garwin, Philip Gordon, Michael Green, Richard Haass, Robert Haffa, Robert Hale, John Hamre, William Inglee, Frederick Kagan, Nina Kamp, William Kaufmann, Steve Kosiak, Andrew Krepinevich, Richard Kugler, Eric Labs, Michael Levi, James Lindsay, Frances Lussier, Thomas Lynch, William Lynn, Jack Mayer, Thomas McNaugher, Michael Meese, James Miller, Derek Mitchell, Mike Mochizuki, Satoshi Morimoto, David Mosher, William Myers, Sam Nunn, Carlos Pascual, Bill Perry, David Petraeus, Lane Pierrot, Kenneth Pollack, Barry Posen, Jack Pritchard, Peter Rodman, Scott Sagan, Stephen Sargeant, Jeremy Shapiro, Eric Shinseki, Peter Singer, Anne-Marie Slaughter, Thomas Stefanick, James Steinberg, John Steinbruner, Strobe Talbott, Richard Ullman, Robert Vilhauer, Frank von Hippel, Stephen Walt, and Dov Zakheim. I would like to dedicate this book to Cathy, Grace, and Lily.

THE SCIENCE OF WAR

INTRODUCTION

SUN TZE'S ANCIENT WORK, *The Art of War*, is a classic that remains as timeless today as when he wrote it several centuries before Christ in China. Questions of morale, leadership, cunning, and innovative tactics are still central to warfare and always will be. Similarly, the Prussian general and scholar Carl von Clausewitz's book *On War*, written two hundred years ago, remains brilliant in its depiction of war as an extension of politics, a fundamentally human endeavor in which national and individual will and the core character of fighting men (and now women) are central in understanding battle and determining outcomes.

But there is also a science of war—that is, a structured, analytical, often quantitative, often rather technical side to preparing for combat. The science of war is also important for keeping peace. It can help improve and ensure deterrence, by scrupulously evaluating the capacities of one's own military and trying to strengthen it where possible. Finally, it is important for defense budgeting and resource allocation, a matter of importance not only to war planners but to all participants in the public policy process.

The use of quantitative tools in defense policy analysis is almost always imprecise, not only in the quality of the data available, but even in knowing what concept and what formula to apply to a given problem. All the tools developed and discussed in this book need to be viewed in this spirit.

We have little choice, however, but to try to refine the science of war as much as possible. What would the alternative be? To base defense budget levels on pure guesswork or politics? To make war plans only using the intuition of generals (or secretaries of defense?). To develop any weapon that seems technically within reach without regard to its likely cost, effectiveness, or other strategic effects? As imprecise as the science of war may be, we must attempt to understand it. And even policy generalists must grapple with it themselves, unless they wish to cede some or all of the defense policy debate to specialists—which cannot be in the national interest in any country. Given the importance of military debates, which in the United States presently involve more than $600 billion a year in annual spending and two significant wars simultaneously, it is essential that as many as possible feel capable of wading into defense debates with some degree of understanding and some sophistication.

The yin and yang of this book is that, recognizing scientific methods in defense analysis to be imprecise, we must nonetheless strive to understand, improve, and employ them. Studying the science of war should never be seen as a substitute for studying the art, history, and contemporary aspects of warfare. It is instead an essential complement.

This textbook is written with the national security agenda and problems of the United States of America central in mind. However, most of the analytical methods discussed here have a wider applicability as well. It is not first and foremost a discussion of contemporary defense debates, however, nor an examination of defense policy. It is about methodology for military analysis. That said, many defense policy matters are ultimately discussed by way of illustrating the use of various methods. From questions of how to size and shape the Pentagon budget, of which wars to fight and how many casualties they might entail, of how large forces should be for peacekeeping and counterinsurgency missions, the methods here should be of considerable use. This book is a textbook, but with a policy edge and a policy motivation.

After a very brief primer on defense matters, specifically concerning the U.S. military, in the following pages of this chapter, the book begins with a discussion of budgeting methods. It focuses particularly on the American defense budget debate. This chapter will probably be of interest to the widest audience. Subsequent chapters on modeling warfare, on understanding defense logistics, and on understanding scientific issues in defense policy such as space warfare, missile defense, and nuclear weapons development are intended for a somewhat more specialized audience—though my goal is always to make the material as straightforward and accessible as possible. Graduate students in public policy should be able to navigate the book's material even if not specialists in the field; undergraduates with sufficient background and interest in national security matters should be able to do so as well.

I have attempted to help draw the material together with several exercises at the end of each of the book's four main chapters. These are designed in the spirit of textbook review and practice questions. Defense analysis being as much an art form as a quantitative exercise or an exact science, the answers provided are not always the only way to tackle the question posed. They should help define a problem, orient a reader and a user regarding the methodologies discussed here, and show one way to begin to attack each problem. They should not be interpreted as the only "right answer" to any question.

Large numbers of people work (in official capacities) all the time on the defense analysis problems considered here. Take budgeting, for example. Specialists in this activity include uniformed officers slaving away within their respective service headquarters, a mix of officers and civilians doing the same sort of work within the Office of the Secretary of Defense, and contractors working for the Department of Defense (DoD) at one of many consulting firms. Combat modeling is quite frequently done by contractors as well (some with strong ties to one service, others with more ecumenical client bases, such as the Institute for Defense Analyses). It is also carried out by the Pentagon's Joint Staff and its regional military commands, and by its war colleges: Quantico for the Marines, as well as the Army War College in Carlisle, PA, the Air War College in Montgomery, AL, and the Naval War College in Newport, RI. All tend to employ elaborate and classified computer models requiring dozens or hundreds of data inputs, generally involving tens of thousands of lines of computer code.

A third main topic of this book, military logistics, is the focus of U.S. Transportation Command based at Wright-Patterson Air Force Base in Ohio (largely for intercontinental transport to a possible combat theater, and supported by various key facilities and organizations such as Air Mobility Command at Scott Air Force Base in Illinois and Military Sealift Command, headquartered in Washington, D.C.). Many others have key roles in logistics, too, including the military services, the individual units making up their combat force structures, and the larger echelons of organizations such as the corps headquarters of the U.S. Army. These organizations get forces to ports or airfields of embarkation, and then upon arrival in a combat theater determine how to support them. Finally, issues in military technology development are handled by the services' weapons laboratories, which number in the dozens throughout the country (such as the Navy's Patuxent River Naval Air Station in Maryland, Dahlgren Surface Warfare Center facility in Virginia, Kirtland Air Force Base in Albuquerque, and China Lake Naval Air Weapons Station in California), by dedicated parts of major defense contractor organizations, and at a more basic scientific level by many of America's universities (such as the Applied Physics Laboratory run by Johns Hopkins University in Maryland, parts of the Department of Energy's laboratories at Los Alamos and Sandia in New Mexico, and Lawrence Livermore National Laboratory in California).

This book does not seek to replicate the detailed calculations used in these various organizations, which together employ tens of thousands of individuals. Nor could it. To some extent, this means that the methods discussed here must be somewhat simpler and less precise than those used officially. However, for many, if not most, purposes in defense analysis, methodological simplicity is not a weakness and may even be an advantage. Too many elaborate models focus attention on detailed numerical analysis—data crunching—rather than on linking the mathematics to a conceptual understanding of warfare and other military operations. Once analysts get into data crunching mode, it is often difficult to keep asking basic—and crucial—questions about one's assumptions. Given this, as discussed in the following pages, official Pentagon models and computations often wind up no more accurate than simpler approaches. Such analyses are also much more difficult for outside organizations (often acting individually or in small groups and without classified information) to mimic. So by necessity, but also by design and by preference, this book seeks simpler methods. In fact, many individuals working within the defense community will themselves use these simpler methods as checks on their more elaborate work, or as a way of orienting themselves and understanding the broad contours of a problem before delving into detail. In defense analysis, a field often afflicted by mistaken assumptions and by a false sense of confidence on the part of its practitioners, simpler is not necessarily less valuable—and it may not even be less rigorous or accurate.

A Brief Orientation to Defense Matters

This book is not a primer on the American military, other countries' armed forces, or defense matters in general. Those interested in probing such issues in detail might consult books such as James Dunnigan's *How to Make War*. Though dated a bit, it is still quite informative. They might also value various Department of Defense publications available at www.defenselink.mil, such as the 2008 National Defense Strategy, the 2006 Quadrennial Defense Review, or more technical and data-rich reports on matters such as personnel statistics and annual budgets. Nor is this a book on military doctrine and tactics. Such subjects are addressed, for example, in the Army's *Operations* manual and in more specialized reports on subjects such as peacekeeping and stability operations. This is also not a book about military culture or about many service-specific issues and interests (see publications such as the *Marine Corps Gazette*, *Airpower*

Journal, *Parameters*, and *Proceedings* for the latter, or more generally Carl Builder's timeless book, *Masks of War*). I am not particularly well equipped to write the latter types of tomes, and in any case, doing so is not my purpose here.

Before delving into methodology and calculations in the pages that follow, a few paragraphs may help situate the general reader who is not familiar with day-to-day defense matters. Most of this book does not require its reader to have a detailed background on the military. That said, since I do not devote much time or attention to explaining "Defense 101" in the chapters that follow, it is appropriate to include a few brief words here, particularly on how the American armed forces are configured, commanded, and operated.

The U.S. military is organized into four independent services: the Army, Air Force, Navy, and Marine Corps. For administrative purposes, the Marine Corps functions within the broader Department of the Navy (meaning there are three "departments," but four services). The services are constitutionally and legally responsible for raising, training, and equipping the men and women of the U.S. armed forces. Together, they currently employ about 1.5 million active-duty troops, with the Army being the largest service (just over half a million active-duty soldiers), and the Marine Corps the smallest (about 200,000 active-duty Marines). Over 200,000 of these 1.5 million are officers; the rest are enlisted personnel. The armed forces also include roughly one million reservists spread across six organizations (each service has a reserve component, and the Army and Air Force also have National Guard units distributed across the country). Military personnel are located at hundreds of bases around the United States. In addition, some 400,000 are presently stationed or deployed abroad, with about half in the broader Middle East / Central Asia theater (as of this writing in late 2008) and the other half distributed mostly in Europe and East Asia.

Since the National Security Act of 1947, the four services have been organized within the Department of Defense, which is run by the Secretary of Defense, a key member of the president's cabinet. A number of entities make up the Office of the Secretary of Defense (OSD), with numerous undersecretaries, assistant secretaries, and other civilian officials directing them. The military services all have their own organizations and units to provide analytical support as well (civilian service secretaries also provide leadership to them). It is the services and OSD, in conjunction with the president's Office of Management and Budget, that work with Congress on the annual budget process. The Pentagon's funding is approved year to year—that is, it is a "discretionary" account, in federal budget parlance. At roughly half a trillion dollars a year (not counting war costs), it constitutes about half of all discretionary spending, approximately one-fifth of total federal spending in normal times.

Since the Goldwater–Nichols Act of 1986, the military's joint-service organizations and operations have also been strong. Within the Pentagon is a joint staff. The four service chiefs, plus a chairman and vice chairman, make up the Joint Chiefs of Staff (JCS). There are six regional military "combatant" commands, again organized and run at the joint-service level (even if only one person from one service can command any of them at a time). These are Central Command, Pacific Command, European Command, Southern Command, Northern Command, and Africa Command (the last two both having been created during the Bush 43 presidency). These commands run operations in their respective theaters, but depend on the services—and thus on Washington—for the means to do so, since they have few if any dedicated forces "of their own." The chain of command to these commanders runs from the President through the Secretary of Defense (SecDef), but not through the Joint Chiefs (though the latter are charged with providing their best military advice on various issues to the President, the SecDef, and the Congress).

There are also four functional combatant commands—Strategic Command, Special Operations Command, Transportation Command, and Joint Forces Command. Each of these ten regional and functional commands is run by a four-star officer (a general or admiral), and each of the service chiefs as well as their vice chiefs is a four-star officer as well. Together these eighteen individuals represent almost half of all the four-star officers in the U.S. military at any one time; most of the others run individual commands within the various services, such as Air Combat Command.

It is useful to be aware of a few more key pieces to the organizational configuration. Numerous joint-service agencies perform tasks of importance across the Department of Defense. They include such organizations as the Missile Defense Agency or, during the recent wars in Iraq and Afghanistan, the Joint Improvised Explosive Device Defeat Organization (JIEDDO, to give an example of the acronyms of which the Pentagon is so fond). They are staffed by military personnel as well as full-time civilian employees of the Department of Defense, who number nearly 700,000. Defense contractors are critical to America's defense efforts as well. Including those working in the defense industry, and in various support activities, their numbers are cumulatively comparable to those in the direct employ of the Department of Defense.

Much of the nation's intelligence community, estimated to cost nearly $50 billion a year by unclassified sources, is funded within the Department of Defense budget, including the CIA and the National Security Agency. Other large chunks of the intelligence community work for the Department directly as well, since in addition to the Defense Intelligence Agency, each of the services and regional military commands has its own intelligence organization.

Much of the military can be organized in an "order of battle"—a detailed force structure typically consisting of several categories of units within each service. Each unit is itself individually numbered and often named as well. Much of this type of information, on not only American but other countries' militaries, can be found in an extremely useful reference guide, *The Military Balance*, published annually by the International Institute for Strategic Studies in London.

The complexity of the U.S. military can be daunting. But as with the overall philosophy of this book's methodologies in the chapters that follow, it is often possible to get a rough feel for the basics and understand the core concepts without knowing every detail or nuance. That said, for those wishing such details, many references can be found to provide them.

CHAPTER I

Defense Budgeting and Resource Allocation

DOES THE UNITED STATES spend too much on defense? Why does its current military budget, even excluding costs for Iraq and Afghanistan, exceed the Cold War average in real-dollar terms, and exceed the budgets of either China or Russia by a ratio of some five to one? If the budget is bloated, how can it be prudently scaled back? Regardless of whether it is excessive, how can proposals for new types of military capabilities be properly evaluated? Only by understanding the components of the defense budget can these questions be seriously addressed.

Before getting into more specific arguments, it is worth noting comments frequently made about America's defense budget—usually by those wishing to make the defense budget seem high or low. Many who wish to defend the magnitude of Pentagon spending often point out that in recent decades its share of the nation's economy is modest by historical standards. During the 1960s, national defense spending was typically 8 to 9 percent of gross domestic product (GDP); in the 1970s, it began at around 8 percent and declined to just under 5 percent of GDP; during the Reagan buildup of the 1980s, it reached 6 percent of GDP before declining somewhat as the Cold War ended. In the 1990s, it started at roughly 5 percent and wound up around 3 percent. During the first Bush term, the figure reached 4.0 percent by 2005 and stayed there through 2007; it grew towards 4.5 percent by 2009. Seen in this light, current levels (including wartime supplemental budgets) seem relatively moderate, if hardly low.[1]

By contrast, those who criticize the Pentagon budget often note that it constitutes almost half of the aggregate global military spending (to be precise, 41 percent in 2006, according to the estimates of the International Institute for Strategic Studies).[2] Or they note that estimated 2009 national security discretionary spending of some $670 billion exceeds the Cold War inflation-adjusted spending average of $450 billion (expressed in 2009 dollars, as are all costs in this chapter) by almost 50 percent once war costs are included (and exceeds the Cold War average modestly even *without* war costs). Or they note that it dwarfs the size of America's diplomatic, foreign assistance, and homeland security spending levels (roughly $13 billion, $25 billion, and $44 billion, respectively, in 2009).[3]

These observations are all simultaneously true, and as such they are probably inconclusive in the aggregate. The U.S. defense budget is and will remain large relative to budgets of other countries, other federal agencies, and even other periods in American history. Yet at the same time, it is modest as a fraction of the nation's economy at least in comparison with the Cold War era. As such, while informative at one level, these observations are of little ultimate utility in framing defense policy choices for the future. We must look deeper.

An initial conclusion of this book is that broad arguments about the adequacy or excess of American defense spending, based on overall observations about the size of the defense budget in a historical, international, or economic perspective, are analytically suspect. In my judgment, it is not possible to reach a definitive judgment about whether the United States overspends on defense through the methodologies of defense analysis or the science of studying warfare. And it is especially difficult to do so using arguments, analogies, and comparisons that invoke the overall size of the defense budget. Only by looking more carefully at how defense dollars are spent can we decide if the budget is excessive (or insufficient); the key is to try to identify missions that are not needed, or weapons modernization plans that are too fast and indiscriminate, or war plans that are excessively cautious and conservative. Again, to offer my own perspective, a number of proposed or ongoing defense programs are quite debatable, potentially unnecessary, and/or wasteful. Pentagon belt tightening is feasible in a number of areas, based on reasoned analytical assessments of America's defense needs. (I am not focusing primarily on the costs of ongoing wars in Iraq and Afghanistan in this specific discussion; those operations are clearly debatable on their own merits. My focus here is on the underlying and enduring American defense establishment.) This book is designed in part to help people identify, study, and debate such possible cutbacks. Then again, because so many of the Pentagon's costs are rising faster than the rate of inflation, economizing in a number of areas may do more to slow the growth of real defense spending than to allow actual cuts.

This chapter first explains different ways the defense budget can be categorized, broken down, and defined. It presents several tools for understanding the budget in greater detail—which is necessary for calculations that seek to understand the fiscal implications of any proposed changes to military force structure, weapons procurement plans, or military operations. One approach, developed largely by former Pentagon official and Brookings and MIT scholar William Kaufmann, is essentially "top-down"—it starts with the total defense budget, and then subdivides it into various pieces. Another approach, used in the executive branch, the Congressional Budget Office (CBO), and elsewhere, takes a "bottom-up" approach. Costs per unit are estimated directly, based on what the unit in question comprises in terms of people, equipment, and training time. Largely because Kaufmann's method is more comprehensively and clearly presented, it forms the basis for my approach—but CBO's observations are used as well, and the numbers presented are thus modified from Kaufmann's original versions (as well as being updated for inflation and cost growth).

After presenting these methods, the chapter then turns to the question of military readiness. Readiness is a major strategic issue, a hot-button political subject, and also effectively the defense activity that consumes at least half the defense budget (since it includes spending on people, as well as their training and other operational necessities). Finally, addressing matters of frequent political and policy discussion, the chapter places U.S. defense spending in a global context, with a particular eye on how American spending stacks up against that of China.

Understanding the Defense Budget: Basic Categories

In discussing the defense budget, it is critical to start with a clear use of terminology. Words can be very confusing in the field of defense analysis.

By official definitions, the U.S. national security budget does NOT capture all major government activities that influence American security, since it includes neither diplomacy, nor foreign assistance, nor Department of Homeland Security operations. It does, however, include the Department of Defense, the intelligence community, and the Department of Energy's nuclear-weapons-related activities.

The national security budget is officially known as the 050 function in the federal budget. International affairs programs, including diplomacy and foreign assistance, are labeled as the 150 account; veterans' benefits are found within the 700 function; and homeland security is distributed among a range of accounts.

It is important to know the distinction between budget authority and outlays. Outlays amount to actual spending—checks written by the Treasury and cashed by individuals or corporations. Budget authority by contrast is a new legal power to enter into contracts, granted by the Congress through appropriations bills. It can loosely be thought of as putting money into the Pentagon's hands (or the Pentagon's account at the U.S. Treasury), which is then gradually spent over the ensuing months and years as contracts are signed and products delivered. Authorized budgets slated for a given fiscal year may be spent very quickly, as with salaries for troops, or slowly, as with contracts for aircraft carriers or other large types of equipment (which take years to build and are fully paid for only when complete).[4]

A given year's national security budget authority tends to exceed outlays when defense budgets are rising, and to fall below outlays when budgets are declining. (For example, if defense budgets have been going up—a given year's budget authority might be $500 billion, but actual spending or outlays for that year might be $480 billion, for example—it is the latter figure that then enters into federal deficit calculations.) The reason is that outlays, as noted, show a lag effect because they reflect the effects of the previous year's budgets, as well as the current year's. There can also be differences, to the tune of several billion dollars a year, between discretionary budgets and overall budgets (the latter count mandatory spending as well). Mandatory accounts do not require annual budget action from the Congress (entitlements like Social Security and Medicare are the most important examples of mandatory spending in the federal budget). However, almost all defense spending is discretionary, meaning no funds are available in a given year unless explicitly appropriated for that year by law.

Unlike other federal agencies, the Department of Defense routinely provides Congress and the public with detailed information about its longer-term plans each time it submits a budget request. As part of its so-called future years defense program (FYDP), it projects roughly five years into the future. During the Rumsfeld tenure at the Pentagon, such information was less forthcoming than had been the norm, but even then the available information was considerable. The Department of Defense also publishes unclassified information on its major weapons programs through documents such as selected acquisition reports (SARs). These show total cost projections and planned numbers of equipment purchases for several dozen of the Pentagon's largest acquisition programs over an even longer time horizon (extending as far out as the relevant acquisition program is expected to continue).

A brief word is in order about the way the Pentagon budget is constructed, which centers around the so-called PPBE process (for planning, programming, budgeting, and execution system). As explained by retired Colonel M. Thomas Davis, the Pentagon begins its efforts to create a budget proposal for Congress with reference to the National Security Strategy (which is itself coordinated and published by the National Security Council, generally under the signature of the President). The Joint Staff at the Pentagon then writes a National Military Strategy based on the National Security Strategy. This is used in the first, or planning, phase of the PPBE process to create documents known as Guidance for Development of the Force and Joint Programming Guidance (formerly done via the Defense Planning Guidance document), which prioritize certain military missions and begin to translate broad strategy into more concrete decisions.

However, it is in the second, or programming, phase where specifics really emerge. The services and defense agencies take the lead at drafting program objective memoranda (POMs) during this phase, subject to the all-important budget ceilings provided by the Office of Management and Budget (OMB). The POMs are then vetted and revised by the Office of the Secretary of Defense. Once so edited, they are used to produce the Future Years Defense Program (FYDP), as noted earlier, a half-decade-long budget plan. Even greater detail is provided about the first two years of that FYDP in the budget request submitted to Congress (however, Congress only appropriates funds for the first of the two years, once it sees the budget proposals and acts on them).[5]

At present, the Department of Energy's share of national security or 050 spending is about $18 billion a year. Another $6 billion is spent here and there by other agencies, but the bulk of the money is for the Department of Defense (DoD).

As for the DoD, estimated spending in 2009 is roughly $650 billion. Viewed in terms of budget authority, the underlying "peacetime" budget for that year in the administration's request was $515 billion. Another $70 billion was requested as an initial "supplemental" for the final months of calendar 2008 (that is, the initial months of fiscal year 2009, prior to the inauguration of the new president). Supplemental budget authority had totaled about $170 billion in 2007 and $195 billion in 2008 (again, in 2009 dollars).

The following section first explains the so-called peacetime budget, breaking it down by category and function, and then discusses supplemental budgets such as those presently used to fund the Iraq and Afghanistan wars.

Breakdowns of the Department of Defense Budget

The U.S. military, in its official budget documents, breaks down its overall budget several ways. Two basic methods show spending by appropriations "title" and by military service (neither includes Department of Energy [DoE] nuclear costs). The service-by-service approach merges Marine Corps costs within the Navy's (since technically the Marine Corps is a separate service but is not a separate department). Many intelligence community costs are found within the Air Force budget when costs are subdivided this way, since it launches and operates many of the rockets and satellites used in intelligence as well as many of the aerial assets.

Another method, devised by former Secretary of Defense Robert McNamara, subdivides spending by what he called military "programs." Rather than allocate the defense budget based on military service, or type of activity, he sought to use broad categories defined in terms of the overall functions they served. They include strategic nuclear capabilities, main combat forces, transportation assets, administrative and related support activities, National Guard and reserve forces, and intelligence, as well as several smaller areas of expenditure. This method is not completely accurate. Many military forces are usable for both nuclear and conventional operations, for example. Should they be viewed as strategic nuclear capabilities or main combat forces? And how to allocate expenditure for equipment first bought for active forces but later transferred to the reserves? Moreover, these categories are broad enough that, even if accurate, they may have only a modest bearing on informing policy choice. Nevertheless, they still give at least an order-of-magnitude sense about how different types of military objectives or main activities translate into costs.

Table 1.1, 1.2, and 1.3 reflect the Bush administration's request for 2009, which is slightly different but not enormously different from what Congress approved (the following does not count any war costs). The budget authority figures include about $2.9 billion in so-called mandatory spending but are predominantly discretionary in nature.[6]

Available documentation, found at www.budget.mil and updated in fresh documents each February with the Pentagon's budget request, provides a great deal of detail for those wishing it. For example, within the military personnel accounts, information can be found on the costs of officer pay versus enlisted pay, on salaries versus accrual financing for the future retirement of current troops, of active versus reserve compensation, and of travel and moving allowances, to name but a few subcategories. Within the procurement budgets, out of which most equipment purchases

TABLE 1.1
DoD 2009 Budget Authority Request by Title (Billions of Dollars)

Military Personnel	128.9
Operations and Maintenance	180.4
Procurement	104.2
Research, Development, Testing, and Evaluation	79.6
Military Construction and Family Housing	24.4
Management Funds, Transfers, Receipts	0.8
TOTAL	518.3

Sources: Under Secretary of Defense (Comptroller), *Military Personnel Programs (M-1), Operation and Maintenance Programs (O-1), Department of Defense Budget, Fiscal Year 2009* (Washington, D.C.: Department of Defense, February 2008), pp. 18, 20; Under Secretary of Defense (Comptroller), *Construction Programs (C-1), Department of Defense Budget, Fiscal Year 2009* (Washington, D.C.: Department of Defense, February 2008), p. iv; *Procurement Programs (P-1), Department of Defense Budget, Fiscal Year 2009* (Washington, D.C.: Department of Defense, February 2008), p. II; and *RDT&E Programs (R-1), Department of Defense Budget, Fiscal Year 2009* (Washington, D.C.: Department of Defense, February 2008), p. II.

TABLE 1.2
DoD 2009 Budget Authority Request by Service (Billions of Dollars)

Army	139.0
Navy	149.0
Air Force	143.7
DoD-wide	86.6
TOTAL	518.3

Source: Under Secretary of Defense (Comptroller) Tina W. Jonas, "Fiscal Year 2009 Budget Request: Summary Justification," Department of Defense, Washington, D.C., February 4, 2008, p. 8.

TABLE I.3
DoD 2009 Budget Request by Program (Total Obligational Authority, Billions of Dollars)

Strategic Forces	9.9
General Purpose Forces	201.9
Command, Control, Communications, Intelligence, and Space	77.6
Mobility Forces	13.5
Guard and Reserve Forces	38.4
Research and Development	52.8
Central Supply and Maintenance	22.0
Training, Medical, and Other	70.6
Administration	18.8
Support of Other Nations	2.2
Special Operations Forces	9.0
Other	0.1
TOTAL	516.8

Sources: See Under Secretary of Defense (Comptroller), *National Defense Budget Estimates for FY 2008*, pp. 1–2, 81; Office of Management and Budget, *Budget of the U.S. Government, Fiscal Year 2008: Historical Tables* (Washington, D.C.: Government Printing Office, 2007), pp. 89, 164; and Allen Schick, *The Federal Budget: Politics, Policy, Process* (Washington, D.C.: Brookings, 2007), p. 57.

Note: Here, the figures add up to a slightly different total because what is presented is total obligational authority, not budget authority. The difference in these two concepts is quite small for our purposes, but it has to do with the possibility that some funds and obligations can carry over from one year to the next (or lapse or be eliminated before being obligated), creating a slight difference between budget authority and total obligational authority in any given year.

Another detail worth noting here concerns the distinction between discretionary budgets and overall total budgets. Discretionary funds must be appropriated each year by Congress. Overall budgets also include mandatory programs and spending, which do not require yearly attention (entitlements are the largest example of mandatory programs in the federal budget). Almost all military spending is discretionary. Mandatory accounts can be positive or negative since they can involve trust funds, user-fee programs, and the like. For example, in 2008 the administration's request for all DoD funding was $643.7 billion, while the discretionary request was for $647.2 billion, meaning the mandatory funding request was "negative."

are funded, subcategories exist for aircraft, vehicles, ammunition, missiles, and other groupings of assets by military service.

Which category is most useful for understanding a given policy challenge or framing a given policy choice depends on the issue at hand. For example, imagine someone wishes to know if the country should move to a smaller but more mobile military. One notional way to create this might be to double the budget for U.S. mobility forces, using savings from a smaller combat force structure to fund the expanded transportation programs. McNamara's categories shown in Table 1.3 would give some sense of how much of the combat force structure might have to be cut to make this possible. The same table could help one estimate how much deep cuts in nuclear programs might save. (In doing so, one would have to remember that DoE nuclear warhead costs, and many missile defense costs, are not captured in the strategic forces category used here. It is also important to know that, historically, nuclear-related costs were far higher than today, typically reaching $80 billion a year in 2009 dollars and perhaps even more by some estimates.)[7] Or if the question was how a 5 percent across-the-board increase in military compensation would affect the defense budget, the information in Table 1.1 could come in handy (incidentally, civilian pay for DoD employees, which totals a bit more than $60 billion a year at present, is located within the Operation and Maintenance budget).[8]

Generally, however, more refined budgetary tools are typically needed for these and other purposes. The previously mentioned tools are informative but not that powerful analytically, nor that accurate. In subsequent sections, we will therefore develop finer tools of defense budget analysis to allow more detailed assessments. First, however, we shall examine the question of war costs and supplemental appropriations.

The Wartime Supplementals

Through fiscal year 2008, Congress had appropriated about $750 billion for military expenses in the broadly defined global war on terror, including $35 billion for Iraqi and Afghani security forces, plus another $40 billion for diplomatic activities and foreign aid for the broader effort. (By comparison, in 2009 dollars, the Korean War cost about $480 billion, Vietnam about $680 billion, and Desert Storm about $90 billion, with 90 percent of the latter costs paid by U.S. allies.) In rough terms, total funding for Iraq approached $600 billion by late calendar year 2008, Afghanistan $150 billion, and DoD homeland security efforts $50 billion.[9]

The preceding figures refer to the additional costs of various military missions. These are the costs above and beyond those already incorporated for the forces in question in the standard defense budget (such as regular salaries and training costs). These incremental costs per troop for current military operations have been very high—over $500,000 a year on average. This was more than twice what had been projected back in 2002. Even once costs for activities such as training Iraqi security forces are removed, costs per U.S. troop deployed have exceeded $400,000 a year.

This was not the first time the Pentagon failed to accurately predict the cost per troop in a major operation (note that this has nothing to do with how long the mission lasts nor how many troops are deployed, but rather the annual cost per troop deployed). DoD has been notorious for its failure to understand deployment costs accurately in the recent past. For example, in the Bosnia mission, initial estimates for the cost of deploying 20,000 troops to the region for a year were $1.5 billion to $2 billion, but actual costs were at least twice the upper bound (in other words, at least $200,000 per deployed troop per year).[10] This is a good case study in how even sophisticated defense analysis tools are often crude, and thus how generalists armed only with fairly simple methodologies can often be nearly as accurate as Pentagon planners.

Why are these numbers so much higher, on a per-troop basis, than those from past wars or from earlier estimates for this very war? It is one thing to have uncertainty in projections of war costs based on not being sure how long a war will last or how hard the fighting will be. (For example, prior to Operation Desert Storm in 1991, the Congressional Budget Office [CBO] estimated that war costs could be roughly $45 billion to $140 billion, when translated into 2009 dollars; actual costs were about $95 billion in 2009 dollars, with just under 90 percent paid by friends and allies.)[11] It is something else to be off by 100 percent or more when the duration and troop strength of a given mission are not in doubt. Numerous reasons help explain this. Military facilities have been developed to be high-tech, comfortable, and useful over a sustained period. Equipment has been worn down by intense operations and damaged by enemy action far more than predicted. New types of equipment to handle specific challenges of recent wars, notably the Mine Resistant Ambush Protected (MRAP) vehicles, have been developed and produced in large numbers. Contractors have been hired to support operations in very large numbers. And quite unabashedly, DoD has added a number of costs for activities not strictly related to the war—such as restructuring its Army brigades—in supplemental requests since about 2005.[12] Accordingly, supplemental procurement funding averaged only about $10 billion annually through 2005, but rose to $27 billion in 2006 and $54 billion in 2007—with a whopping $74 billion requested for 2008 (again, expressed in 2009 dollars).[13]

Understanding the Defense Budget: Kaufmann Methods

While the Pentagon's basic budget categories are helpful, those summarized earlier do not suffice when one seeks to analyze policy alternatives. The breakdowns by service and title fail to fully describe the missions to which defense dollars are devoted. Even the McNamara program elements fail to give any insight into the force structures designed to carry out those missions or the costs associated with them on a per-unit basis.

In recognition of this dilemma, longstanding Pentagon advisor and Brookings scholar, the late William Kaufmann, created two additional breakdowns of force structure, weapons purchases, and related Pentagon costs that can complement the McNamara method and in some cases provide more useful analytical tools. One approach subdivided costs by geographic region of the world, the other by main combat formations of each of the military services. The second is especially useful.

Implicit in the Kaufmann methodology, by the way, is the notion that to save money in the Department of Defense budget, one probably needs to cut real capabilities. That view is in contrast to the hopes of many that a savvier way of buying weapons, or more restrictions on "revolving door" practices between defense industry and the military services that have sometimes fostered corruption, or privatization of more defense support activities will lead to big savings without painful cuts in military muscle. While such savings are sometimes attainable, and while all these policy issues are worthy of attention in their own right to foster good governance, experience suggests that they are unlikely to yield big windfalls of savings that can then be redirected. Perhaps the best summary of how to think about such savings comes from a report written by former Congressional Budget Office Assistant Director and current Pentagon Comptroller Robert Hale. Aptly titled, "Promoting Efficiency in the Department of Defense: Keep Trying, But Be Realistic," the report suggested that creative and effective defense reforms might save hundreds of millions or even a few billion a year if done well. This is real money, to be sure, but it is a far cry from the tens of billions believed achievable by some in the past. (For example, some Pentagon critics have noticed the 20 to 30 percent savings that resulted when DoD outsourced certain totally nonmilitary activities like paycheck processing and assumed that such monies could be found more generally throughout the budget. That has generally proven quite hard to achieve on a broader scale in the normal peacetime defense budget.)[14]

By contrast, there is something of a consensus among analysts that private contractors can save DoD money in overseas operations, and perhaps even more than 20 to 30 percent, with some estimates approaching 50 percent. Such savings are achievable partly due to lower per-person costs when contractors are able to hire lower-wage non-Americans, and partly due to the fact that contractors need not be paid when operations end (whereas comparable numbers of uniformed personnel would presumably be retained in the force structure even after an operation was over). However, this benefit must be weighed against the risk that contractors may not be quickly available, may not be willing or able to operate in dangerous settings, and may not be sufficiently well regulated and disciplined to be compatible with some missions requiring great care in the use of force. In addition, American contractors are generally not individually cheaper than government employees, according to a recent study by the Office of the Director of National Intelligence, and in fact may be at least 50 percent more expensive per year.[15] On balance, contractors do have an important role for the U.S. military. But this fact is already being exploited, since in Iraq the United States is using one contractor for every uniformed man or woman—2.5 times the highest rate of any previous large conflict, relative to the number of deployed troops.[16]

Kaufmann's Geographic Breakdown of American Defense Costs

Kaufmann's geographic approach subdivided America's military missions by overseas theater—Europe, the Atlantic sea lanes, the Far East, the Persian Gulf, Latin America, and Africa. Most combat formations were assigned accordingly, though some were attributed to American territorial defense, or to missions such as nuclear deterrence and intelligence. Kaufmann created a taxonomy that, like McNamara's, had about ten main categories. Again as with McNamara's, the allocations were constructed such that their sum total equaled the aggregate defense budget.

The basic logic of Kaufmann's allocation scheme is simple and appealing. And if accurate, it can allow for a clear method of assessing the fiscal implications of various American security commitments, such as protection of Persian Gulf oil or of key allies like Japan, Korea, or the countries of Western Europe. But is it right?

While not lacking merit, Kaufmann's geographically oriented defense budget breakdown is probably the most controversial of the major methods considered here. Some military assets are indeed designed primarily for one type of operation in just one or two places—such as frigates designed to protect ships as they traversed the Atlantic and Pacific during the Cold War, or aircraft carriers and Marine Expeditionary Units routinely deployed to specific regions like the Persian Gulf and Western Pacific. In these cases, it is not hard to apportion costs on a regional basis, at least roughly speaking. (For ships and ship-based Marines, examining their homeports and typical deployment patterns can further help determine if they should be assigned primarily an Atlantic/Mediterranean or a Pacific/Indian Ocean orientation.) During the Cold War, Europe was the primary intended heavy combat theater for air-ground operations, making it logical to attribute the costs of most Army and tactical Air Force units to that region (at least until the Vietnam War upset the assumptions, and the calculus). After the Cold War, the focus for warfighting planning shifted to the Persian Gulf and Korean peninsula, as Pentagon documents such as the bottom-up review explicitly noted. So this method has a certain logic to it.

Kaufmann's last breakdown was done in 1992, when the Pentagon still worried about a possible Russian resurgence and the resulting hypothetical danger to countries like the Baltic states. As such, he estimated that a substantial fraction of the overall defense budget was for the defense of Europe. It is not clear if he would reach the same conclusion today, though in the aftermath of Russia's August 2008 invasion of Georgia, arguably one could. (His amounts are shown in Table 1.4, displayed as percents of the overall Department of Defense budget. Only the budgets for nuclear forces as well as national communications and intelligence are not divvied up by region.) The share of the defense budget focused on the Western Pacific, with an eye not only towards North Korea but also the rise of China, and the more general ascendance of that part of the world in economic and strategic terms, conceivably might change if Kaufmann were to redo his estimates today. Kaufmann's figures were quite modest; a larger estimate might be appropriate now. In addition, the United States has also been more active in Africa in the years since Kaufmann last wrote, beginning with the ill-fated Somalia mission but also including refugee relief in Central Africa, counterterrorism cooperation with countries in the Sahel region, and now the creation of Africa Command (AFRICOM). But Kaufmann's numbers nonetheless do reflect his initial assumptions about the state of the post–Cold War and post-Soviet world, and as such still have some relevance today.

TABLE 1.4
Kaufmann Estimates of DoD's Spending by Geographic Region under the "Base Force" of the First Bush Administration (in Percent)

Strategic Nuclear Deterrence	15
Tactical Nuclear Deterrence	1
National Intelligence and Communications	6
Northern Norway/Europe	5
Central Europe	29
Mediterranean	2
Atlantic Sea Lanes	7
Pacific Sea Lanes	5
Middle East and Persian Gulf	20
South Korea	6
Panama and Caribbean	1
United States	3

Source: William W. Kaufmann, *Assessing the Base Force: How Much Is Too Much?* (Washington, D.C.: Brookings, 1992), p. 3.

The resulting budget tools Kaufmann created are most useful when trying to assess how much the country spends defending various overseas interests and allies—and how much it might save if it reduced certain commitments (or how much it might add if it increased commitments). This framework can help in debates over allied military burdensharing, in discussions of how much the United States spends defending Persian Gulf oil, and so forth.

However, it would be a mistake to take this framework too literally, because it is not really how the Pentagon plans its forces or budgets. It is very rare to have a combat formation that could only be usable in one part of the world. To be sure, some headquarters capabilities, some planning staffs, and some intelligence assets are devoted to a specific region.

Also, the occasional combat unit, such as the Army's 2nd infantry division in Korea, is sometimes very closely associated with one region. But most American combat forces are flexible, as they must be. The United States has too many global allies and interests to create, in effect, separate force structures to defend each. The cost of doing so would be prohibitive.

Most American forces are based in the United States and are deployable to whatever region national command authorities might need to send them. Even formations thought of as devoted primarily to one place may, when a crisis erupts, wind up going elsewhere. The Army drew down large numbers of European-focused forces to fight in Vietnam (not to mention Desert Storm in 1991), and one of two brigades previously based in Korea was sent to Central Command's operations in the Iraq/Afghanistan theater in recent years (after which it returned to the United States, not Korea). Speaking of Afghanistan, it is a place for which no combat units had been designated prior to 2001. Looking to hypothetical scenarios in the future, there are no dedicated American forces for addressing instability in Pakistan, peacekeeping in Kashmir, humanitarian relief in Africa or South or Southeast Asia, or a range of other possibilities. Yet capacity must be retained for addressing one or more such scenarios at a time, even if each is individually unlikely to occur.

As such, Kaufmann's framework, while useful, is more notional than precise. The methodology by which he makes his regional cost estimates is also not explained in any detail in his writings. Kaufmann's personal reputation for rigor and great knowledge in the field make it likely that his estimates are as reasonable as any other—given the inherent limitations of tackling the problem in this basic way. But they are not easily reproducible or confirmable.

A corollary of this observation is that, even if a given overseas interest of the United States disappeared, or was deemed no longer requiring or meriting American military protection, the resulting implied decline in the defense budget would generally be less than Kaufmann's method suggests. This is because some of the forces he allocated to a given region are also important for other regions—if not in a primary role, then at least as a strategic reserve.

Costs of Individual Combat Units

An additional methodology developed by Kaufmann breaks down the overall defense budget according to the main combat force structure—Army and Marine Corps divisions (active and reserve), special operations forces, Air Force and Marine Corps tactical air wings, Navy carrier battle groups and amphibious ships and sea control forces, airlift, sealift, prepositioning, strategic nuclear forces including submarines, bombers, land-based ICBMs, missile and air defenses, early-warning assets, and national intelligence and communications systems. Again, as with his geographic allocation scheme, Kaufmann's goal was to create a meaningful yet simple tool, so he sought to avoid excessive complexity. His approach was to assign all DoD costs to these categories so they added up to exactly the overall size of the defense budget.

Then Kaufmann went further, making his approach the most analytically useful of all those considered in this chapter to date. Specifically, he divided the cost for a given category of combat capability by the number of units within that category to provide a cost per unit.

In fact, dozens of forces are designed to support combat formations, especially in the Army with its multiplicity of units and subunits focused on various aspects of logistics, intelligence, administration, air defense, other specialized weaponry, and the like. Kaufmann's logic, however, was that in the end these capabilities exist only to support combat formations. As such, associating a certain proportionate number of support capabilities with each main combat formation, and allocating costs accordingly, is a reasonable way to create a good framework for policy choices. It allows a rough answer to the question of how much the country would save (or expend) if, for example, we cut (or add) an Army division or Air Force wing to the military force structure.

This approach of Kaufmann's is extremely useful. And it does a much better job of capturing the real cost implications of many policy alternatives than do most official Pentagon numbers. For example, if the Department of Defense were asked how much it could save by cutting an Army division out of its force structure, it might give (and often has given) a very modest figure. Army divisions only have about 16,000 to 18,000 soldiers in them in general, and in making an estimate of savings, the Army would probably focus on personnel and operating costs (not equipment), since the former are often the only guaranteed expenses for a given military formation in a given year. However, at least 40,000 people are actually associated with a given Army division once support costs are included, and over time the Army will save a certain amount in reduced equipment expenditures when reducing force structure (even if savings in a given year cannot be specified exactly). So Kaufmann's method would often show a cost about five times greater for a unit such as an Army formation—and there is little doubt Kaufmann's estimate would be more accurate in general, at least in indicating potential savings in the medium to long term.

That said, the Kaufmann approach has its limitations. It is not entirely accurate. Not all defense costs can be proportionately allocated to main combat formations. Some military costs such as the research and development of new technology, strategic intelligence operations, and combat headquarters operations are to a large extent independent of the size of the force structure. Thus, cutting the number of combat units does not reduce the former costs proportionately, if at all. Yet Kaufmann's method would often allocate such costs proportionately to combat force structure, implying greater savings than reality would allow. As for the costs of new weaponry, reducing purchases by a given percentage (as implied by cuts to force structure) generally would not allow fully commensurate savings, since economies of scale would be lost as fewer weapons were purchased.[17] This effect can be mitigated by encouraging defense companies to consolidate as defense budgets decline, among other means. But that process takes time, and at some point the benefits to the Pentagon and taxpayer of having multiple firms compete for the DoD's business can be partially lost. In other words, reducing the total purchase of a given type of weapon by say 10 percent does not typically produce savings as great as 10 percent, yet Kaufmann assumes it would.

Kaufmann's figures may overestimate actual savings from any force structure cuts by 20 to 30 percent, roughly speaking. To see why, we can examine Army force structure as one example. It has historically been divided into three relatively equal groups—combat units, deployable support units, and nondeployable or institutional units. Most support units can be cut proportionately when main combat forces are reduced, but the link between combat forces and institutional capabilities is much less direct.[18] So Kaufmann assumes comparable savings in all three types of costs, when in fact major savings are likely to prove possible only in two of the three. Despite the limitations of Kaufmann's methods, they are typically the most accurate way to get back-of-the-envelope estimates.

In his 1992 book, Kaufmann made the following estimates of the average annual costs of different types of main combat units. These were projections for what those costs would be in the mid-to-late 1990s (expressed here in billions of constant 2009 dollars; Kaufmann had used 1993 dollars, and I multiplied by 1.53 to make the conversion). They include yearly personnel and operating costs, as well as an annualized estimate of how much investment a given unit requires to be outfitted with weaponry and other equipment. (In another part of the book, Kaufmann estimates differences in cost from one type of division or air wing to another, but he himself works with the average numbers shown in Table 1.5 for comparative purposes.)[19]

Kaufmann's approach was to multiply each of these unit costs by the number of that type of combat unit in the force structure, add $2.8 billion for special forces, as well as $30.5 billion for national intelligence and communications, and $60.8 billion for nuclear forces ($60.2 billion of that being for strategic forces, just $0.6 billion for tactical forces)—and wind up with the total defense budget exactly. In other words, he assumed that all other types of assets and units in the U.S. military force structure could be associated proportionately and directly with main combat forces, as noted earlier. This assumption is incorrect in a technical sense, yet is hugely useful as an approximation, if one's main goal is to have a rough gauge of how force structure translates into costs.

Although the preceding figures have all been converted into 2009 dollars, one could argue that they are not really quite ready for use in the present day.[20] In fact, defense costs typically rise substantially faster than inflation (as discussed later in the section on the Congressional Budget Office's methodology). This is largely due to the growing complexity of military technology, which accounts for most cost growth from generation to generation of equipment such as fighter aircraft.[21] These considerations would suggest that the preceding figures should all be larger than shown, by as much as 50 percent or so—necessitating one more round of modifications of these numbers to make them usable today.[22]

However, I have chosen not to modify Kaufmann's numbers in this way. The main reason is that as noted, Kaufmann's cost figures back in 1992 actually overestimated the savings that could be obtained by cutting unit x or unit y (and overestimated as well the costs of adding force structure). Kaufmann assumed implicitly that research and development costs could be cut proportionately when force structure is reduced. He did the same for supporting capabilities—reconnaissance units, electronic warfare units, refueling aircraft, air defense and artillery units, and so on, not to mention headquarters and administrative/managerial oversight, and the service's educational institutions, health care systems, and the like. By his logic, such capabilities would not exist absent combat forces; they exist only to support combat forces, and if combat units are reduced that should imply commensurate reductions in support capabilities. Kaufmann's method helps focus the mind on the potential for overall savings if efficiencies are pursued throughout the Department of Defense when main combat force structure is reduced, but they are nonetheless too high, because certain

TABLE 1.5
Kaufmann Estimates for Average Annual Costs of Key Combat Formations (Converted from 1993 to 2009 Dollars, Billions)

Active Army Division	5.2
Active Marine Division	4.0
Reserve Army Division	1.0
Reserve Marine Division	0.9
Active Army Brigade Combat Team	1.2
Reserve Army Brigade Combat Team	0.2
Active Marine Tactical Air Wing	3.5
Active Air Force Tactical Air Wing	3.0
Reserve Marine Tactical Air Wing	0.9
Reserve Air Force Tactical Air Wing	0.60
Navy Carrier Battle Group	6.4
Amphibious Ship	0.15
Sea Control Ship	0.2
Airlift Aircraft	0.02
Sealift or Prepositioning Ship	0.02

Source: William W. Kaufmann, *Assessing the Base Force: How Much Is Too Much?* (Washington, D.C.: Brookings, 1992), pp. 70, 74, 82.

Note: I have taken the liberty of modifying Kaufmann's estimate for an active Marine Corps division from $2.6 billion to $4.0 billion because this is the one estimate of his that seems somewhat incongruous with other data and estimates in this chart. I have also taken the liberty of estimating how much a brigade combat team—increasingly the standard unit of organization and measure for the U.S. Army—would cost.

activities like research and development and central oversight cannot be reduced in lockstep with changes in force structure. By my best estimate, as noted 20 to 30 percent of the costs Kaufmann associated with a given unit would be very hard to pare back in the event of marginal changes to the force structure.

On balance, these imprecisions tend to balance and cancel out each other. As such, Kaufmann's figures are probably the most reasonable and accessible starting points available, once adjusted, as I have explained earlier. Once all is said and done, they are probably within 10 to 20 percent of more accurate (but generally elusive) estimates in most cases.

Understanding the Budget: CBO

The Congressional Budget Office, staffed with several dozen individuals specializing in defense budgeting yet generally writing unclassified documents for a wide audience, has considerable expertise in the field, combined with the goals of achieving simplicity and clarity in its analysis. That combination makes it especially useful in this discussion.

CBO has generally been reluctant to provide comprehensive information on the costs of main combat units. So Kaufmann's numbers, as adjusted in the previous pages, are perhaps our best guide here. But CBO has offered cost estimates over the years for a number of weapons systems. In Table 1.6, costs are expressed in millions of 2009 dollars and equipment is assumed to have a lifetime of thirty years (except fighter and attack aircraft, which are assumed to have a lifetime of twenty years, and Virginia-class subs, which are assumed to last thirty-three years).[23]

CBO has also produced a different type of overall defense budget analysis regularly over the years. Looking at the whole DoD budget, it projects the medium-term costs of the Pentagon's plans. In these studies, CBO analyzes information more by title—operations and maintenance, military personnel, military procurement, research and development and testing and evaluation—rather than by specific combat unit.

To carry out these projections, CBO takes the existing budget. It then adjusts it for several factors: planned changes to the force structure, known policy changes such as pay raises for personnel or purchases of new weapons, and Pentagon plans for reconfiguring bases or support organizations. It also offers its own best estimates of what is politely termed "cost growth"—the degree to which actual costs will exceed predicted ones. This is a major concern in many types of Pentagon budgetary accounts. For example, over recent decades, operations and maintenance costs on a

TABLE 1.6
CBO Cost Estimates (Millions of 2009 Dollars)

System	Annual O&S Costs	Procurement	Annual Average
DD-963 Spruance Destroyer	41	450	56
DDG-51 Arleigh Burke Destroyer	31	1,500	81
CG-47 Ticonderoga Cruiser	42	1,700	98
Virginia Class SSN Attack Sub	41	2,600	120
A-10 "Warthog" Aircraft (1 Plane)	5	15	6.0
F-15 Eagle	7	65	10
F-15E Strike Eagle	7	78	11
F-16 Falcon	6	32	8.0
F-14 Tomcat	10	78	14
F-18 Hornet	8	62	11
F-18E/F Super Hornet	8	86	12
F-22 Raptor	10 (est.)	150	18
F-35 Lightning II/AF variant	7 (est.)	89	11
F-35 Lightning II/Marine variant	8 (est.)	96	13
F-35 Lightning II/Navy variant	10 (est.)	110	16
MX Missile	4.1	160	9.3
Minuteman Missile	2.4	75	5.0
B-52 Bomber	16	250	25
B-1 Bomber	23	310	33

(*continued*)

TABLE 1.6 (*cont.*)

System	Annual O&S Costs	Procurement	Annual Average
B-2 Bomber	40	850	68
Trident Submarine (plus 16 D5 missiles)	70	2,200 (+750)	170

Sources: David A. Fulghum and Amy Butler, "Passing the Buck," *Aviation Week and Space Technology*, February 11, 2008, pp. 24–25; Caitlin Harrington, "Congress Warned Over Additional Raptor Buy," *Jane's Defence Weekly*, February 13, 2008, p. 14; Raymond Hall, David Mosher, and Michael O'Hanlon, *The START Treaty and Beyond* (Washington, D.C.: Congressional Budget Office, 1991), pp. 139–40; Congressional Budget Office, "Total Quantities and Unit Procurement Cost Tables, 1974–1995," Congressional Budget Office, Washington, D.C., 1995, pp. A-17 through A-22; Lane Pierrot, *A Look at Tomorrow's Tactical Air Forces* (Washington, D.C.: Congressional Budget Office, 1997), pp. 3–5; and Eric J. Labs, *Transforming the Navy's Surface Combatant Force* (Washington, D.C.: Congressional Budget Office, 2003), p. 11.

Note: Most numbers here are limited to two significant figures to avoid implying greater precision than available data warrant.

per-troop basis have grown about 3 percent annually in real terms—even when they were expected to hold roughly steady. The rate of growth varies, but it is always greater than inflation, and it has a fairly steady trajectory to it. In recent decades, health care and environmental cleanup costs associated with base closures have been among the largest drivers of increased operations and maintenance (O&M) expenditures, with the former growing from about $15 billion a year to almost twice that from roughly 1988 through 2003, and now reaching $40 billion. Aging weaponry, by contrast, has apparently had only a modest impact to date.[24] While it is not a given that such cost-growth trends will continue in the future, absent a radical change in management methods they probably will.

For weaponry, CBO makes use of detailed information about the normal growth of equipment costs as a given system moves through development into production. A detailed literature is available on the subject, created by defense research organizations such as the Institute for Defense Analyses, RAND, and others. Room for judgment always exists about the degree to which future weapons will display different cost-growth patterns than past ones, meaning debate often arises about CBO's numbers.

But to first approximation, assuming cost growth in the future like that which has occurred before is usually a more accurate method of predicting future costs than relying on the DoD's generally optimistic official figures.

Weapons typically cost more to acquire than the Pentagon initially anticipates. For example, space vehicles as well as Army ground combat vehicles both tend to cost 70 percent more to develop, and ground vehicles also typically cost 70 percent more to produce, than initially estimated. On the other end of the spectrum, ships typically cost about 16 percent more to develop than first projected, and 11 percent more to produce (after adjusting for any changes in numbers of weapons bought). Using such information, CBO can crudely estimate cost growth for each weapon based on where it stands in its overall development and production cycle. On average, weapons cost around 20 to 50 percent more to acquire than initially forecast.[25]

CBO also has weapons-specific models, most notably for fighter aircraft, that predict cost based on weapons performance characteristics. For fighter aircraft, cost estimates are based on aircraft weight, maximum speed, the amount of advanced materials in the airframe, the number of contractors involved in the effort, the technological maturity of software, the maximum airflow through the engines, engine thrust-to-weight ratios, specific fuel consumption, and avionics. The basic concept is to look at previous aircraft, take their performance characteristics for all these key attributes, run a regression to create a mathematical formula, and then plug in the expected performance features of the new plane to calculate expected costs.[26]

In its costing methodology, CBO also needs to find a way to estimate procurement costs for small equipment. Cumulatively, such weaponry constitutes roughly 40 percent of all DoD investment spending—that is, the sum of procurement plus research, development, testing, and evaluation.[27] But documentation for such smaller equipment is not detailed enough to allow CBO to do a system-by-system assessment as it does for large weaponry. As such, once it has projected cost growth for large weapons using detailed Pentagon documentation (including selected acquisition reports), it then makes an estimate for minor systems on the assumption that they will continue to account for 40 percent of total procurement spending.

CBO often underscores the uncertainty in budget estimates by creating lower and upper bounds on its projections. The former are typically based on Pentagon expectations about costs, the latter on models and historical experience. For example, CBO recently calculated a higher bound for long-term DoD budgets that was about 13 percent higher than its lower-bound estimate. The average over the next two decades for the higher bound would be roughly $605 billion, versus $535 billion a year, expressed in constant 2009 dollars. These figures do not count major war costs or Department of Energy nuclear weapons expenditures. Of the $70 billion in possible cost growth, about $40 billion would be in the operations and support accounts, the remaining $30 billion in the investment accounts.[28]

Military Readiness

Most defense dollars go to one of two main activities in the United States: preparing forces for combat and other missions over the longer term, through research and development and procurement of weaponry as well as other efforts, and ensuring the near-term combat readiness of forces for missions in the here and now. In other words, the former activities are primarily (though not exclusively) about investment in hardware, the latter more about people and training. The former are addressed in the chapter of this book on military technology; the latter here.

Military readiness refers to prompt and immediate response capability for plausible missions. It is defined by the Joint Chiefs of Staff as the ability of the armed forces to deploy quickly and perform initially in any military contingencies as they were designed to do.[29]

The question of military readiness is in large measure a budgetary question, since many readiness gauges can be substantially influenced by how defense resources are allocated. It is also a critically important matter in many national security debates and in many American political debates as well.

Indeed, readiness has often been a political football. In the 1970s, America's military was alleged—in large part correctly—to have gone "hollow." A substantial force structure existed, but it was not particularly strong within, and it did not hold up very well when called upon to perform. During the mid-to-late 1990s, similar allegations about the U.S. military were made as the armed forces were downsized in the aftermath of the Cold War.[30] In the first decade of the twenty-first century, as the Iraq and Afghanistan missions have continued relentlessly, many have spoken of a U.S. Army that has become broken.

What are these criticisms about? What do terms like "hollow" and "broken" even mean in the context of military preparedness?

It is important to differentiate military readiness from other important criteria for assessing military preparedness. In fact, readiness for assigned missions is only one concern that strategists and policymakers must emphasize.

Policymakers also need to determine if they are worrying about the right possible missions. Since many wars historically come as surprises to countries fighting them, this is a very difficult and crucially important task. Being ready for a war or other important mission that does not happen is no great solace if a country's armed forces prove unready for what they are ultimately asked to do.

Planners also need to worry about future threats. (This is sometimes called "long-term readiness," though that is a confusing use of terminology best avoided in favor of a better term like investment or modernization.) Long-term planning can compete with the near-term ability to carry out key functions—or readiness. Emphasizing the latter requires lots of money and time for training, for spare parts and fuel and ammunition, and for focus on threats or military missions that are already recognized as important. Emphasizing the longer-term, by contrast, tends to require more money and time devoted to research and development, to professional military education, and to experimentation with new warfighting concepts and technologies.[31] Dollars spent on R&D cannot be spent on fuel or ammunition; officers' time devoted to running future-oriented exercises and experiments is not available for drilling forces for near-term missions; units trying out new concepts are not as able to practice on more immediate tasks or potential tasks. So it is important to think in terms of tradeoffs.

Indeed, readiness for one mission can also compete with readiness for another. The Balkans wars were seen as distractions from "real" warfighting priorities and readiness requirements in the 1990s, since formal war plans were more focused on Northeast Asia and the Persian Gulf. The Afghanistan war was not anticipated or planned very well before it occurred, in part because of the preoccupation with what were considered more immediate threats requiring that readiness efforts be focused on them. The U.S. military prepared much better for the invasion phase of the Iraq War than for the post-invasion phase, suggesting that readiness considerations had emphasized classic invasion scenarios at the expense of counterinsurgency and stabilization missions. Also, in the last few years the ledger has shifted, with the constant preparation for ongoing operations in Iraq and Afghanistan dominating the time and attention of soldiers and Marines to the detriment of training for other missions.

Even once one defines readiness fairly narrowly and leaves broader strategic issues for separate debate and analysis, challenges remain in assessing it. The military is always short certain capabilities as defined by its doctrine and its process for determining military "requirements." How much does one shortfall matter, in a given area of personnel or equipment or training, when other capabilities and other resources are robust? Just as with the debate over metrics in Iraq, it is always possible to find readiness metrics that make a predetermined political case—that seem to suggest that the military is either in fine shape or is falling apart. It is for such reasons that Richard Betts, the Columbia University professor, subtitled one chapter in his seminal book on the subject of readiness "lies, damn lies, and readiness statistics."[32]

It is not particularly helpful when the Pentagon limits its public discussion of readiness metrics to a few categories that it selects, especially when it fails to present a historically significant period of time to provide perspective on these categories. For example, in its 2009 budget request, the Pentagon showed data for just six categories of readiness, and in each case only showed two years' worth of data.[33]

Yet we must not throw up our arms in frustration. It is possible to think systematically enough about readiness to shed at least some light on the question of how well prepared the American military (or another military) may be for the likely missions it is designed to conduct. Otherwise, we are left to resorting to the selective use of statistics and anecdotes, which is not an effective way of determining how to allocate military resources.

The most dependable way to evaluate readiness is to fight real enemies, since war is always a major learning experience, and a great judge of capability. However, this is not a reliable or desirable way to assess readiness. The downsides are obvious—the possibility of suffering a major defeat before learning what went wrong, not to mention the inherent costs in blood and treasure. However, it is still true that the United States employs its military often enough that real operations are significant sources of information and important barometers of preparedness.

Consider just a few examples from the American experience of the last three decades. In 1980, the United States failed to rescue hostages in Iran, and lost eight servicemen when an accident during refueling in the Iranian desert led to the cancellation of the mission. Poor coordination among the different military services involved and a lack of realistic training may have contributed to the tragedy. In Lebanon in 1983, the U.S. military learned about the importance of force protection and the difficulty of maintaining neutrality during a peacekeeping mission when the Marine barracks were bombed. In short, it found out it was not ready to protect its forces against the types of threats that presented themselves on that battlefield (which, of course, foreshadowed the types of suicide truck bombings that became tragically commonplace in the ensuing quarter century).

The Kosovo War in 1999 was impressive in many ways, and debunked the notion that the U.S. military had somehow been rendered hollow by its post–Cold War downsizing. But it also revealed how immobile the Army was when it tried to send a modest number of Apache helicopters to nearby Albania. The Iraq War required major learning—or perhaps the relearning of past lessons since forgotten—about counterinsurgency operations, with the U.S. military more impressive from 2007 on than in the earlier years of the war.[34]

More happily, in the 1986 bombing of Libya, as well as the 1989 invasion of Panama, Ronald Reagan's military buildup was at least partially vindicated as U.S. forces (and equipment) performed impressively. In the Panama case, improvements in military command and control instituted in the 1986 Goldwater–Nichols act helped also ensure a well-directed mission.[35]

But clearly, waiting for the next war to evaluate a new readiness initiative or take stock of what traditional efforts have accomplished is not a preferred, or reliable, way to evaluate readiness. As such, other more mundane methods must be employed. They typically begin with dividing the subject into three broad categories: personnel, training, and equipment. Within each of these three umbrella groupings, one can then break things down further by looking at the different subcategories of troops and various types of military specialists, assessing the capabilities of each key subunit. One can do similar things when evaluating training for various units, as well as the availability and condition of equipment—not only in regards to major vehicles but also spare parts and ammunition.

Three challenges stand out when performing such accounting tasks, which are in and of themselves generally straightforward (if somewhat tedious and labor-intensive). First, has the military established proper benchmarks for readiness within each area? If standards for training or personnel are too low, for example, reaching those standards in a given month or year may not ensure true readiness. The error can work in the other direction as well. If a certain amount of ammunition is deemed necessary by the Pentagon, any shortfall relative to that requirement then becomes, or at least appears to be, a readiness problem even when it may not be. A case in point was the U.S. cruise missile inventory several years ago—consisting of almost ten times the number of cruise missiles used in Operation Desert Storm, yet somewhat below what the DoD "requirement" specified.[36] A resulting political storm resulted over the alleged failure of the Clinton administration to be properly ready for combat, when in fact the armed forces faced a shortfall only relative to what was perhaps an overly ambitious goal.

Second, and in some ways even harder, how does one determine the relative military importance of any given shortfall? If nine out of ten readiness measures for a given unit are fine, but the last involves a mission-critical resource (such as ammunition for guns), a severe shortfall in one area of readiness could trump the adequacy of other preparations. This type of situation can lead to failing readiness grades for a whole division or air wing (often called the "C Rating") when that unit has only one specific problem. Sometimes that is appropriate, but quite often it is not, because shortages of one type of capability can often be redressed by substituting it with another.

Relatedly, in the late 1990s, two entire Army divisions were deemed "unready" because they had each sent one brigade to the Balkans. In fact, each still had two brigades that were available for other missions.[37]

Even more strangely, consider a case where a unit is supposed to acquire a new type of fighter by a given date. If the deliveries of that fighter are slower than expected, the reported readiness of the unit can actually *decline* even while in the process of getting better equipment. That happens, for example, when the formal readiness accounting standard changes faster than new equipment can be delivered. There is also subjectivity in some of the integrated readiness measures that makes them vulnerable to a certain political influence at times. It might be preferred to emphasize how good readiness is (for those wanting to show a "can do" attitude and document their service's good use of the defense resources it has been given). Alternatively, a given military service or official may emphasize how bad readiness is, relative to some standard (for those wishing additional resources, for example).[38]

The best antidotes to these problems are to present data in a relatively raw form, use consistent standards over time in evaluating it, and avoid sweeping generalizations about "passing" and "failing" readiness grades. Generally, it is best to show statistical trendlines, and use qualitative judgments to supplement these numbers rather than replace them.

A third problem for an outside specialist when assessing readiness is that not all information is easily available. Some is quite appropriately classified; a key shortage in a key type of equipment is not the sort of thing to broadcast (especially if those within the system recognize the problem and address it without outside prodding). Other information that is less sensitive should be shared; it should enter into political debates about properly resourcing the defense budget and wisely allocating funds among competing accounts.

Consider several examples from each area of readiness in 2008. My purpose in providing all the details is not to imply that today's readiness problems are unique or monumental (though some are serious, it is true). Rather, the goal is to use 2008 as a detailed case study to show how different indicators can provide different suggestions about the severity of a readiness problem, necessitating some effort to integrate the indices when seeking an overall evaluation.

Personnel

Key measures of personnel readiness include the experience and aptitude of typical troops, the availability of individuals with critical specialized skills, and the ability of the military to recruit new members as well as retain those already in the service. Many of these can be measured. Admittedly, some key readiness metrics, such as the aptitude and creativity of flag officers, cannot be easily understood by peacetime gauges of preparation. However, data and analysis can help assess most types of readiness trends.

In recent times, there has been considerable concern about a lowering of personnel standards. It is worth reflecting on this situation in some detail as an illustration of the types of considerations that arise in such debates. One general rule is that there are always strains on the military, shortfalls in its equipment, skill sets lacking among its people, and specific weaknesses in near-term combat training. However, sometimes these weaknesses add up to a serious problem and sometimes they do not. Hence, each individual readiness statistic or measure needs to be placed in perspective.

Consider several trends in U.S. military personnel that arose during the George W. Bush administration. The military accepted more recruits with general equivalency degrees rather than high school diplomas; it enlisted a higher percentage of applicants scoring very low on its aptitude tests; and it also took on more individuals over forty years old as first-time military personnel than before.

The trendlines on age and G.E.D. degrees have begun to cause considerable concern in the last year or two—though the nature of the problem needs to be kept in perspective. The G.E.D. is considered academically equivalent to a high school diploma, and certainly the military can ensure that anyone with such qualifications is up to par by testing them in other ways, too. Moreover, as of 2005, 90 percent of DoD recruits continued to have high school diplomas, comparable to the 1985 figure at the height of the Reagan buildup. And the typical recruit scored better on the Armed Forces Qualification Test (AFQT) in 2005 than in 1985. That said, while figures for the other services have remained good, the Army has experienced some problems, with the high school graduation figure for 2005 dipping to about 80 percent (worse than the 1985 figure of 86 percent, though still much better than the typical level of around 55 percent in the 1970s).[39] By 2007, the percentage of high school graduates had declined to below 80 percent; it rebounded modestly to 83 percent in 2008, but further progress is still needed.[40]

There has been a recent rumor that West Point graduates have been leaving the service at drastically increased rates as soon as their minimal obligations are satisfied. In fact, this appears not to be true. The last year for which data is available as of this writing (the class of 2002, which was eligible to leave the service as of 2007), showed a 68 percent reenlistment rate, only 4 percentage points below the 1990s average.[41] More generally, company-grade officers (first and second lieutenants as well as captains) have not been leaving the force at a greater than normal rate either; the average rate during the Iraq War has been less than the average rate of the late 1990s, for example.[42] A similar conclusion is true of majors.[43] Nonetheless, the Army is now short several thousand officers in aggregate.[44] This is largely because the Army is increasing the number of officers needed as it enlarges the number of brigades in its force structure. In addition, the Army did not retain enough young officers in the early 1990s, meaning the current pool of officers from which to recruit for mid-level positions is too small.[45]

Other matters are more worrisome as well. While the number of individuals scoring relatively high on aptitude tests remains better than the 1980s, trends are in the wrong direction.[46] Moral waivers for matters such as criminal history have increased substantially in recent years, with a total of 860 soldiers and Marines requiring waivers from convictions for felony crimes in 2007, up by 400 just from the year 2006. While most of the convictions were for juvenile theft, and the aggregate total is modest compared with the size of the force, only by arresting such trends will the quality of the force be ensured.[47]

Divorce rates have leveled off somewhat at about 3.5 percent, after reaching 3.9 percent in 2004, and are not worse than in the general population—but they are still above the 2.9 percent level of 2003.[48] Suicide rates are a significant problem for the military, and an extremely tragic reality for so many troops and their families. The rate reached 17.3 per 100,000 soldiers in the U.S. Army in 2006. That is not far off from the age-adjusted and gender-adjusted average for the U.S. population on the whole (for males, for example, the rate is 17.6 per 100,000), but rates are at record levels today—much higher, for example, than the rate of 9.1 per 100,000 soldiers in 2001—and as such a serious reason for worry.[49]

For one group of soldiers surveyed in 2008, among those who had been to Iraq on three or four separate tours, the fraction displaying signs of post-traumatic stress disorders was 27 percent (in contrast to 12 percent after one tour and 18.5 percent after two). As of early 2008, among the 513,000 active-duty soldiers who have served in Iraq, over 197,000 had served more than once, and over 53,000 had deployed three or more times so many are now at such risk.[50]

Some of the problems mentioned here for the Army are hardly without precedent. For example, in the late 1990s the Air Force pilot shortage was more than 10 percent, as no-fly zones over Iraq taxed airmen and airwomen's patience, and high pay rates in the civilian economy enticed many to doff their uniforms. This would clearly have been a problem for any operation requiring a large fraction of the force structure, such as a major war. Through careful monitoring of the deployment burdens of various units and individuals, however, it was at least temporarily manageable for the challenges of the 1990s. In addition, certain other military specialties were overused, and various individuals were overdeployed, because the force structure had not been properly constructed for frequent deployments that were especially demanding of a particular type of military skill set. Shortages of certain "low density / high demand" military specialties also arose. These specialties existed within the military in only modest numbers, but were frequently needed in certain kinds of operations, and included skill sets such as computing, aviation, foreign languages, policing, and other military activities that could become particularly (and unexpectedly) important in a given mission. Individuals with such skills could often find better opportunities in the private sector. It can be hard to anticipate shortfalls in such areas without access to more detailed information on military manpower than most analysts are generally able to access, especially since the Department of Defense is not always as forthcoming as it could be with relevant information.

On top of that, some units were undermanned and had to be reinforced by individuals borrowed from other units when preparing to deploy. This was acceptable on the whole, but led to situations where some people were deployed with a different unit than their own.⁵¹ If their own unit was itself later deployed, the same person might wind up deploying abroad twice in a relatively short time; the military personnel system did not protect them from such inadvertent overuse. This was more a matter of fairness to military personnel, and sustainment of good morale, than a crisis for warfighting readiness, but it was important to address just the same. Again, the moral of the story is that not all problems with military readiness are of equal concern, though most merit attention and remedial action.

In the preceding cases, targeted policy interventions usually occurred. For example, pilot bonuses were increased.⁵² In addition, in some cases a higher level of future attrition was simply assumed to be likely (especially at a time when a strong civilian economy created many high-paying commercial jobs for pilots), and more military pilots were recruited and trained to compensate. Recruiting and advertising budgets, as well as the number of personnel assigned to the recruiting task, can also be varied. In fact, there is a fairly detailed econometric literature on just how well each type of policy tool tends to work historically (in terms of increasing recruiting or retention per dollar spent). Such measures tend to be capable of addressing most discrete, specific problems within two to five years—if that much time is available.

Other changes can be considered as well. More military specialties can be rewarded with differential, added pay (rather than linking compensation so linearly to military rank). Because military pay, while never truly enough to compensate those who actively risk their lives for their country, is nonetheless reasonably good by comparison with private sector jobs for individuals of comparable age, experience, and education in the U.S. population, the idea of selective bonuses for certain specialties would seem reasonable. Equity and fairness considerations need not preclude such an approach of selective targeted raises and bonuses.⁵³

Military pension plans could also be modified. The military retirement system is essentially an "all or nothing" operation. Stay in for twenty years and become fully vested and immediately eligible for full benefits; stay in a day less and receive nothing. This approach probably hurts retention for those considering whether to stay in the military or not as they approach five to ten years of service. That is the period when another few years of military employment promises no accrued pension benefits whatsoever (unless personnel wind up staying the full twenty years), compared to many private sector jobs that would begin to vest them quickly. These types of reforms may be considered simply as a matter of fairness, or of keeping up with the times and the changing nature of the American economy. Yet they may not be given sufficient attention absent a problem in readiness that motivates innovation.[54]

Details of the exact state of the U.S. military have only limited bearing on the purpose of this textbook, of course. But they are provided in some detail to give a sense of the kind of data that are available—and of the kinds of policy challenges that need to be faced. The broad analytical conclusion from the preceding survey is that one always needs to ask how serious a given shortfall might be when measured against the overall capability of the military (and likely demands of war); what policy recourse might be available to address a given problem as well as how long it might take to implement the repair; and what, if any, immediate stopgap solutions can be adopted to mitigate the strains on people in the short term. It is also important to study trends; problems may either get worse or ameliorate with time. Those that are worsening naturally demand the greatest attention.

If an overall assessment was to be offered of the U.S. ground forces in 2008, it is that of an Army and Marine Corps under serious strain but collectively holding up. Most indicators are not worsening, though they are less healthy than in most periods of the 1980s and 1990s. As such, complacency is hardly in order; we should be concerned that the reasonably good readiness of the military today is fragile and not sustainable indefinitely, even if at present it seems to be sustainable for the foreseeable future. Moreover, at the individual level, many soldiers and Marines are facing enormous hardship, raising fairness and equity issues for a democracy at war, and asking so few to do so much for so long on behalf of the nation.[55]

Training

In the Vietnam and immediate post-Vietnam eras, military training was not nearly rigorous enough. However, great attention was focused on this matter in the 1980s, demanding regimens were developed, resources were amply provided for training, and subsequent military performance was seen to improve greatly, as validated not only in training but in certain military operations such as the 1986 air attack on Libya, the invasion of Panama, and eventually Operation Desert Storm. Ever since that period, Reagan-era standards have remained important in determining how forces should be prepared—from basic training to specialized training to unit training at main bases to large-formation training at the various weapons schools and combat training centers.

Challenges remain, however. In normal peacetime, one concerns the growing role of simulators. As flight simulators and even tank simulators become much more realistic, to what extent can they replace the need for true training? Military personnel need to work with real weapons, realistic (if simulated) battlespaces, and real ammunition—not only at a technical level, but to acclimate people to the pressures and fear of combat. That said, do we really know for a fact that tank crews need 800 miles a year of driving their vehicles (rather than 600, with many more hours on simulators)? Or that pilots need twenty to twenty-four hours in the air per month, the late–Cold War norm, rather than today's fourteen to eighteen?[56]

These questions underscore the degree to which tracking readiness cannot be purely a technical or an accounting exercise. Judgment is always needed. Mistakes are always possible, and sometimes they are not appreciated until a force engages in actual combat. One reasonable guideline is that we should be wary about making rapid major changes in how we train, absent periodic validation under stressful realistic circumstances (such as actual wartime operations) that combat skills are remaining strong.

An even more difficult problem has been faced by the Army and Marine Corps during the Iraq and Afghanistan wars. Soldiers and Marines have virtually no time to do anything more than deploy to the theater of combat, return, rest, and then prepare to go back. Generalized training in other types of combat, besides the counterinsurgency and counterterrorism now being carried out in Iraq and Afghanistan, is by necessity being neglected. The assumption is that forces who performed so brilliantly in classic combat in 2003, and who have been so hardened by ongoing combat of a different type since then, will remain proficient for the full range of possible missions for the foreseeable future even without the full range of training as required by official doctrine. But that is an assumption, not an obvious truth. The assumption can be periodically tested by asking modest numbers of troops to be subjected to tough assessments of their skills in other types of combat on the training ranges (even if there is not enough time nor enough resources to do so for most formations). But again, a level of uncertainty exists in measuring readiness that is hard to eliminate entirely.

Equipment

Combat units clearly need vehicles in working order, enough ammunition to fight for a certain time, and enough spare parts to fill projected demands. The question is not just instantaneous readiness, but the ability to sustain operations for some reasonable period.

Assessing equipment readiness is in the end probably easier than doing so for people or even training. It is generally a question of countable hardware, not human skills. One challenge is in evaluating the importance of a shortfall in which equipment inventories drop by a certain moderate amount, say 10 to 15 percent relative to nominal requirements of "mission capable" aircraft or tanks or trucks or ships. At what point does the cohesion of a tank unit—its ability to carry out basic operations such as maneuver and attack in a coordinated fashion—suffer in that case, for example? Put differently, at what point does expected combat performance (number of accurate rounds delivered on target, attrition rates over a certain period of time in battle, and so on) decline precipitously and by a much larger amount than the reduction in available equipment or manpower might suggest? To some extent, the answer to this question will be scenario-dependent of course, making the calculation quite hard in some cases. As a broad rule of thumb, however, the military's own collective judgment and official rating systems tend to suggest that shortfalls in the range of 5 to 15 or 20 percent of most types of equipment are tolerable, but that larger deficits quickly become of serious concern.

Sometimes a shortfall in one area can be balanced out by surpluses elsewhere. Consider the state of the U.S. Army in recent years. For most major types of vehicles—all classes of helicopters, Abrams tanks and Bradley fighting vehicles, medium weight trucks—there has been no major crisis due to the wars in Iraq and Afghanistan. No more than 20 percent of the total inventory of most weapons has been in the Central Command theater at a time (according to Congressional Budget Office data published in 2007). For most major fighting vehicles and helicopters, there was no shortage of usable equipment for forces based back in the United States. There *were*, however, notable shortfalls of up-armored HMMWVs, MRAPS, Strykers, and two of seven types of trucks. For the trucks, since there were substantial surpluses in some of the other five categories of trucks, there was probably little major problem. For the armored vehicles, however, there would clearly be great difficulty in finding a way to deploy many to any new scenario that might develop fast. So the Army equipment readiness issue is quite specific—potentially serious for some scenarios, much less so for others. On balance, while Iraq and Afghanistan have taxed the equipment inventories of the U.S. ground forces in particular, the far greater strain at this point is on people, not weaponry.[57]

America's Budget in International Perspective— and the Question of China's Military Spending

Most of the previously mentioned budgetary tools are focused on the U.S. defense budget to allow evaluation of possible specific policy alternatives. But it is also important to understand America's defense budget in a global context. Comparative military spending levels are hardly definitive for measuring relative power or predicting combat outcomes; that is why a later section of this book focuses on combat simulation and modeling. But budget comparisons are an important and salient gauge of the effort and resources a country puts into its armed forces, if nothing else. While outputs matter more than inputs, the latter are hardly insignificant.

At this moment in history, two broad questions stand out. First, why is American defense spending so high compared with the military budgets of other countries? Second, and somewhat relatedly, how should one interpret the rapid growth in the defense resources of the People's Republic of China?

On the first point, U.S. military expenditures are indeed remarkably high. There is no other way to put it. As noted earlier, according to the most commonly accepted comparative measurement by the International Institute for Strategic Studies in London, U.S. expenditures represented 41 percent of the world's total in 2006. In fact, America's share grew further in the next couple of years, given the high costs of war supplementals.[58]

As noted earlier, in 2009, U.S. defense spending as a fraction of the nation's gross domestic product is estimated at roughly 4.5 percent—hardly a small number, and significantly more than the 3.0 percent total to which defense spending dropped in 2000 before beginning its upward trend. But again, that 4.5 percent figure is about half the norm of the 1950s and 1960s, and about three-fourths the average figure during the Reagan years. U.S. defense spending is also now about 20 percent of federal government outlays, in contrast to nearly half in the 1960s, for example.[59]

Nonetheless, American defense spending is enormous. It easily exceeds the Vietnam and Reagan-era peaks in real-dollar (inflation-adjusted) terms. It is nearly half the world's total military spending, whereas U.S. GDP is only one-quarter of the global aggregate. It dwarfs China's military spending of about $135 billion (see the following pages for more on this), or Russia's spending of about $75 billion (these are 2006 figures, expressed in 2009 dollars). Most of the next tier of top military spenders are American allies (France and Britain each spent about $60 billion in 2006, Japan $45 billion, Germany $42 billion, and Italy $34 billion). Saudi Arabia and South Korea rounded off the top-ten list at $33 billion and $27 billion, respectively, with India next at $25 billion. Then came another slew of American partners and allies including Australia ($19 billion), Brazil ($18 billion), Canada ($17 billion), Spain ($16 billion), Turkey ($13 billion), Israel ($12 billion), the Netherlands ($11 billion), the UAE ($10 billion), and Taiwan ($9 billion). Among major American worries, the lead spenders are Iran ($8 billion), North Korea (about $2 billion to $5 billion, though estimates are difficult to obtain), Venezuela ($3 billion), Cuba ($2 billion), and Syria ($2 billion).[60] All these latter allocations are extremely modest—as are the working budgets of groups such as al-Qaeda and various extremist, overseas militias (which measure in the tens or hundreds of millions of dollars a year at most).[61]

For some, the preceding information is enough to conclude that U.S. defense spending is not only large but exorbitant and unnecessary—especially in an era of large U.S. budget and trade deficits. Indeed, when NATO and East Asian allies are figured in, the Western alliance system accounts for about 70 percent of global military spending, and when other allies in places like South America are also included, as well as countries having security partnerships with the United States such as Taiwan, Israel, and the Persian Gulf sheikdoms, the broader U.S.-led global alliance system's military spending exceeds 80 percent of the world's total.

However, economists are divided about whether U.S. budget and trade deficits are deeply harmful of the American economy. Certainly, most would prefer smaller deficits, but even if that is the case, whether the military should take the lead on producing savings is a debatable proposition. (In my own work, I have used the budget deficit as motivation for seeking more painful reductions in defense modernizations than the services would prefer, but there is no conclusive way to insist that this *must* be done as a simple matter of long-term national security.)

Moreover, simple comparative metrics are hardly adequate to determine proper force sizing or budget levels. There are multiple reasons for this, beginning with the simple observation that American defense spending would look huge comparatively even at $200 billion or $300 billion—and it is hard to find proposals for cutting American spending below $200 billion among serious defense scholars (indeed, it is increasingly hard to find proposals for amounts below $300 billion). This suggests that the comparative approach is more useful for establishing context than reaching conclusions.

An important reason for America's high defense spending is its large number of overseas interests and allies. Allies add to the strength of the Western alliance on what might be called the supply side, but they also potentially add burdens to the United States on the demand side. With several dozen formal security partners, and large military deployments in three main regions of the world (East Asia, Europe, and the broader Middle East) as well as smaller commitments in numerous other places, the United States has many actual and potential military obligations. Moreover, to be reached by American forces, these distant theaters all require a substantial effort from the U.S. military, adding to the difficulty of potential missions—and limiting the utility of comparative defense budget analysis, since potential enemies would generally be fighting on or near their home turf.

Another important explanation for the disproportionately large U.S. defense budget is that the United States seeks a major qualitative advantage in military capability. It is not interested in a "fair fight," that is to say an even competition. Rather, it seeks a major military advantage. Such superiority, so the logic goes, should enhance deterrence by reducing the likelihood that other countries will choose to challenge the U.S. military. To put it more negatively, history is full of examples in which the smaller military, and quite often the less expensive one, prevails in battle. While high spending cannot totally overcome the distinct possibility that the underdog will win in war, it can certainly make the underdog's job much harder.

Qualitative superiority also helps compensate for the modest size of the U.S. armed forces, which have been severely strained in the process of trying to stabilize two mid-sized nations of about 30 million people each in recent years. The active-duty American military, at about 1.5 million uniformed personnel, is certainly not large in a historical or current international perspective. Not only is it down from the range of 2.0 to 2.25 million that characterized the post-Vietnam Cold War military (and much higher levels during Vietnam and Korea), it represents less than 10 percent of the world's total of 24 million individuals under arms. China leads the way at 2.1 million. India at 1.3 million, North Korea at 1.1 million, and Russia at 1.0 million are not too far behind the United States by this measure. (South Korea, Pakistan, Iran, Turkey, Egypt, and Vietnam occupy the next tier in terms of size, with armed forces ranging from 450,000 to 700,000 personnel each.)[62]

The premium that the United States places on operating globally, and with a high-tech advantage, leads among other things to very large budgets for the Navy and Air Force. Each of these has a budget comparable in size to that of the U.S. Army; it is unusual for a country's air and naval capabilities each to cost as much as its ground forces.

The mention of China, moreover, underscores the point that American defense planning must look to the future, not just the present. A country with 2.1 million persons under arms, a military spending level exceeding $100 billion a year and growing fast, and irredentist claims on a key American economic and security partner (Taiwan) demands at least some concern. Yet it is also important to place this spending in perspective and not be more worried about it than the evidence would warrant.

The official defense budget of the People's Republic of China (PRC) at market exchange rates was $47 billion in 2007. But that figure is just the starting point for gauging China's military resource allocations.

The PRC official number substantially understates actual military resources for two main reasons. First, due to limited transparency in its official papers and budgets, it fails to include many defense activities commonly recognized as intrinsic to a military budget by standard NATO definitions (or common sense). These feature foreign arms purchases, military-related research and development, nuclear weapons activities, and paramilitary forces. Altogether, the absence of these categories of expenditure creates a significant error in the official budget. Thankfully, corrections can be made with reasonable accuracy. For example, China's arms purchases from abroad (largely from Russia) in recent years have reached roughly the $2 billion to $3 billion annual level, and its nuclear costs are probably in a roughly comparable range, as are its research and development costs. Various types of industrial subsidies given to the defense industry may total $5 billion to $10 billion a year. Overall, when adjustments are made to the official budget to account for such undercounting, the total budget probably increases $15 billion to $20 billion a year.[63]

Second, as a developing economy, China (like many other countries) has a number of military-related costs that are quite inexpensive by Western standards. When China's defense budget is expressed in dollars, therefore, and a comparative sense of its military resource allocation is sought, it is more useful to express figures in terms of purchasing power parity than straight exchange-rate comparisons. Making such conversions is difficult. While the fact that a Chinese soldier can be fed and housed much more cheaply than a Western soldier is an advantage for the PRC, and should probably be acknowledged in any careful comparison, the fact that a typical Western soldier's education and salary are much more expensive reflects a qualitative advantage of that soldier. So not all costs can be converted the same way.[64]

When all is said and done, the United States government has typically estimated China's actual military resource levels as two to three times its official numbers. A reasonable midpoint estimate for China's spending level in 2006 is thus about $135 billion, and the 2007 level may have reached close to $150 billion.[65]

As noted, this only provides a general context. For a possible war, say over Taiwan, would China's greater geographic proximity and greater sense of national vital interest allow it to fight the United States to a draw—or at least have a chance to do so—in any future war? It is hard to determine the answer from defense budget comparisons alone. Trying to reach judgments on this complex subject will have to await the chapter on modeling. For now, the point is simply to establish some sense of the scale of China's effort and the pace at which it is accelerating that effort.

Illustrative Cost Calculations and Comparisons

To illustrate how the preceding defense budget costing tools can be used, and to address several issues important in their own right as well, it is useful to consider a number of specific questions in which budgetary methods can answer a question—or at least inform a broader debate. They are examples of important questions, and my approaches to addressing them are examples as well; other approaches can be imagined in most cases.

QUESTION 1: What is the most economical way to destroy 100 fixed aimpoints within a given country?

ANSWER: This is a very "nuts and bolts" type of question, less strategic than tactical and even technical, but it is nonetheless an example of the kind of practical matter that force planners must consider frequently.

Of course, this question can have many variants, depending on which weapons are available and capable of carrying out the attack in question. There is no right answer across all scenarios.

To simplify the problem, though, let us focus on one particular choice: whether to use a stealthy, "penetrating" bomber like the B-2 (carrying cheap, "dumb" bombs) to attack targets directly, or whether to use aircraft from standoff distances that do not penetrate the airspace in question, but instead use cruise missiles (such as today's B-52s, or perhaps a military variant of a 747 in the future). Of course, this comparison only makes sense if both the bombers and the cruise missiles can successfully reach their targets. Or, if some losses are expected, the costs of replacing the losses should be factored into the calculation, as should the implications of losing bomber pilots—or having them captured. For that and other reasons, simple cost comparisons like the one done here are only a single element of any assessment about which weapons to employ, and which to purchase and maintain in a country's force structure.

Since the United States already has B-2 and B-52 bombers, this example is somewhat contrived. But it could still be relevant in deciding which type of plane to buy in the future.

To simplify, assume that the B-52 like aircraft can carry ten cruise missiles and the B-2 bomber can carry twenty bombs (either a truly "dumb" bomb with no guidance package, or an inexpensive short-range guided munition such as the joint direct attack munition or JDAM, which is terminally guided by GPS). Further assume that the cruise missiles and bombs are equivalent, in terms of expected effect on the target—their explosive power and their accuracy are comparable, so the same number of weapons must be used on each target regardless of which ordnance is selected. (In fact, assume that two weapons must be dropped on each target.) If the targets must all be destroyed on the first sortie of planes, to maximize the shock value and prevent the enemy from recovery and prompt retaliation, we need a way to deliver 200 weapons to target in a single salvo. Using these assumptions, that would require either ten B-2 bombers or twenty B-52/747 aircraft.

To complete the calculation, some information on costs is needed. From Table 1.6, assume that the marginal production cost of buying B-2s is $850 million an airplane (assuming that the research and development of the plane was already completed—meaning that such costs are "sunk" and as such no longer relevant to future policy choices). And assume the cost of a B-52 or 747 carrying cruise missiles to be $250 million, again not far off from reality. Cruise missiles are assumed to cost $1 million each in the following; JDAM bombs $20,000 apiece.

So buying ten B-2 bombers and 200 JDAMs would cost about $8.505 billion, whereas buying twenty B-52/747 aircraft and 200 cruise missiles would cost $5.2 billion. Clearly, if only one mission of this type were expected over the lifetime of the aircraft, the B-52/747 option would be cheaper—assuming that indeed it would be equally effective militarily. Factoring in estimated operating costs over a thirty-year lifetime (see

Table 1.6), the answer would not change much at all, since the added cost per plane of maintaining the B-2 would be largely canceled out by the fact that more B-52s would be needed to accomplish the same mission according to assumptions about likely weapons payload.

Of course, in reality one would have to reach some conclusion about how many such missions might be expected over the lifetime of the planes, and how many other types of missions might be assigned to the planes, before reaching any firm decisions about the respective budgetary attractiveness of each option. As such, these kinds of cost calculations are best viewed as tools to aid the military judgment process, not means of arriving at a definitive "correct answer" in and of themselves.

QUESTION 2: What is the most cost-efficient way to carry out the forward presence mission of the U.S. attack submarine force?

ANSWER: This is again a fairly specific, technocratic question—but it is needed to inform a broader discussion of where to base U.S. military assets and how to operate them in overseas theaters.

American attack submarines are used frequently for intelligence-gathering and related purposes in places such as the Western Pacific Ocean and the Persian Gulf. Here we consider different ways of carrying out this mission.

The U.S. attack submarine force has had multiple missions over the years, many secret, and the Navy has generally been extremely reluctant to describe how it has sized and shaped that part of its force structure. But in recent years, Eric Labs at the Congressional Budget Office and others have been increasingly successful in gaining information about the employment of the attack submarine force. It now appears that the main mission of much of that force—or at least, the mission that is the most taxing on the Navy, and hence the one that determines the size of the attack submarine fleet—is maintaining presence in theaters such as the Western Pacific and the Persian Gulf/Strait of Hormuz. Understanding the operations of navies such as those of the PRC and Iran, as well as understanding the properties of the waterways near these strategically crucial regions—and being ready for rapid response in the event of crises in such places—have become the main missions that determine the size of the attack submarine (SSN) force.

As such, to give a concrete example of policy choices, consider the following: How much could the Navy save by basing more SSNs at Guam?

Three are homeported there at present, and focused naturally on missions in the Western Pacific. The Navy could reportedly find room for at least another half dozen SSNs in the island's port facilities (once suitably upgraded). Of course, other considerations must enter into the final decision-making as well, ranging from the somewhat mundane matter of the quality of life for Navy crews based at Guam, to the more strategic matter of the relative vulnerability of attack submarines to surprise attack when based on Guam or Hawaii or California. But assuming such considerations do not decidedly argue against Guam, the cost comparison becomes of great interest.

An SSN operating from a mainland U.S. port cannot maintain much time deployed in places where it can carry out reconnaissance, train, and be prepared for crisis response. In fact, its ability to stay on station is surprisingly limited, much less than for surface ships apparently—only about thirty-six days a year. By contrast, an SSN homeported on Guam can do about three times as well (roughly 106 days a year). Employed in this way, it will spend slightly more time per year at sea than a U.S.-based submarine, but manage triple the useful time.

Using the SSN more frequently in this way reduces its expected service life somewhat, from thirty-three to twenty-seven years, and increases annual operating costs by about $3 million, from $41 million to $44 million. In addition, about $10 million in added infrastructure would need to be built for each SSN added to Guam.

As such, the cost comparison can be approximated in this way. One SSN on Guam can do the work of three based in California, according to CBO. With a procurement cost of $2.6 billion per sub, the annual average cost of operating one SSN from Guam is about $140 million (dividing the procurement cost by twenty-seven years and then adding in average annual operating and infrastructure costs, in the same sort of way as employed in Table 1.6 earlier). By contrast, the annualized cost of operating three SSNs from the mainland United States is about $360 million. Clearly, if cost is the main issue, Guam wins the comparison hands down. But again, such calculations are aides to decision makers, not final answers themselves, as many other factors typically must be considered when determining force structure and basing.[66]

Of course, this option could not be pushed too far; not only would Guam run out of space for Pacific-oriented submarines, but placing too many in one place would increase the dangers of a successful enemy surprise attack. Moreover, forward presence is not the only mission of the submarine fleet, so a certain number of submarines are needed for surge purposes in a crisis or conflict regardless of the efficiencies associated with forward homeporting. But for a modest additional number of submarines, Guam basing is an interesting option.[67]

QUESTION 3: How much per year, in peacetime, does the United States spend defending Persian Gulf oil?

ANSWER: Since the Cold War ended, possible military scenarios in the Persian Gulf have typically represented one of the two main sets of possible operations around which the Pentagon has built its combat force structure (the other set being in East Asia). In addition, the Navy and parts of the ground forces (especially the Marines, and some Army light forces) have focused more broadly on maintaining presence and quick-response capability across multiple theaters. That would seem to suggest three chief overall missions for main American combat forces, two geographically specific and one more general.

Given that the current peacetime defense budget of the United States is just over $500 billion, this framework might seem to imply costs of about $100 billion a year to defend the Persian Gulf. That estimate can be reached by first factoring out $200 billion of the defense budget devoted to research and development, intelligence, homeland defense, nuclear forces, and other such costs not easily attributable to any geographic theater, as well as the costs of a core military infrastructure including major commands and educational institutions and the like in the United States (see Table 1.3). Then, the remaining $300 billion can be divided into three relatively equal parts.

This is clearly a very rough way to gauge costs. Another way might invoke the Bill Kaufmann breakdown of U.S. military spending by region. He estimated in 1992 that the United States spent 20 percent of its peacetime defense budget on defense of the Persian Gulf and broader Middle East (see Table 1.4). Applying that same percentage today would also suggest a cost of about $100 billion a year at present (again, not counting the costs of the ongoing Iraq War).

However, these figures may be too high. Many forces that would be assigned to Central Command in wartime are available for other purposes as well. (This is the obverse of the current situation, in which forces that could otherwise be used in Europe or East Asia or elsewhere are taking their turns on deployment to Iraq and Afghanistan.) A figure of $100 billion a year may not be a bad estimate of the costs of forces that would be *most likely* deployed to operations in the Persian Gulf, as their most probable main mission. However, this overlooks the fact that many of them could have secondary purposes as well.

So the real question is how much *less* would the United States spend on its overall military if the Persian Gulf were somehow dropped from the list of overseas commitments and possible wartime theaters? That requires an effort to estimate what force posture the United States would want to keep and thus is inherently somewhat interpretive and subjective. Regardless, however, the savings would be somewhat less than $100 billion a year, since some of the forces normally assignable to the Persian Gulf might need to be kept for other possible scenarios (such as stabilization of a collapsing Pakistan or Indonesia, multilateral enforcement of a peace deal between India and Pakistan over Kashmir, a temporary international trusteeship for an independent state of Palestine, a sustained air patrol in the Taiwan Strait if China/Taiwan tensions heat up again and remain hot over an extended time, or a possible Russian menace to the Baltic states, now NATO members).

This question is too complex to lend itself to a precise answer. But figures in the general range of $50 billion a year are probably reasonable answers to the question posed. Such a figure is probably what should be used when, for example, one wishes to compare the costs of alternative energy policies to the status quo.

QUESTION 4: After current wars are over, should the United States reduce its military readiness to fund transformation?

ANSWER: This question again goes well beyond budgetary matters, but budgetary calculations can help inform the answer to the question.

To be specific, and dramatic, let us assume that half of the entire U.S. military is put into a second-tier readiness status, such that its training activities are scaled back dramatically—perhaps to a level only slightly greater than the normal training and readiness level of the National Guard and Reserve. That might suggest a two-thirds reduction in readiness activities and resources, including those to buy fuel, ammunition, spare parts, and other such supplies.

To estimate savings, we could begin with that part of the operation and maintenance budget focused on training, maintenance, and the like, and reduce it accordingly. (That is, we would exclude those O&M expenses related to civilian pay, military health care, environmental cleanup, and the reserve component; we would also assume that neither military pay nor investment accounts would decline under this option.)

As shown in Table 1.1, the current O&M budget is about $180 billion a year; civilian pay is about $60 billion of that total. Cumulative costs in the O&M budget for health care, environmental cleanup, basic base maintenance, recruiting, and the like, as well as the military reserves and National Guard, total another $60 billion or so. That leaves $60 billion for what might be called core readiness activities for the active forces. The respective amount spent by the half of the force structure in question is about $30 billion. Reducing this by two-thirds would save $20 billion a year.

More detailed answers to more specific proposals, say about reducing the readiness of one service or another, could be calculated by looking at the respective O&M budget of the service in question and breaking it down similarly. Precise answers would obviously require much more information, including detailed information on which types of supplies could be cut back by what percent in light of a given reduction in readiness. Such precise answers are ultimately needed if a policy change is being budgeted for. But they are often so difficult to produce that it is useful, for brainstorming purposes, to have a way of obtaining a general and approximate estimate like that provided here.

QUESTION 5: How much would the United States need to cut back the Navy or Air Force to add two divisions to the Army?

ANSWER: Using Bill Kaufmann's estimates from Table 1.5, we can estimate the annual average cost of an active Army division (with about 16,000 troops in the division, including four brigade combat teams, and 25,000 to 30,000 more soldiers supporting it indirectly) at about $5.2 billion. Providing two more would therefore cost $10.4 billion a year (discounting economies of scale to first approximation, as Kaufmann does). Consultation of Kaufmann's list of the costs of other major formations leads to several possibilities for finding that money. The core of one approach could cut three wings of fighter aircraft (at seventy-two aircraft per wing), or two aircraft carrier battle groups (each with about seventy-two aircraft itself, as well as the carrier and about four major warships as escorts), or some combination thereof.

As noted before, Kaufmann's methods, while extremely useful as simplifications and approximations, have their limits. They overestimate savings when cutting forces and overestimate costs when expanding them (the reason for the latter is that some fixed or semi-fixed costs like R&D, central oversight, intelligence, professional educational institutions, and early expensive parts of weapons production runs need not be repeated—or at least need not be incurred at earlier cost rates—when additional forces are added). But other cost imprecisions tend to balance these out. So likely errors are in the range of 10 to 20 percent.

It is important to note that the previously mentioned costs are average annual costs, smoothed out over the entire lifetime of the typical weapon in a given unit. Because of political and budgetary realities, however, planners are usually worried less about costs averaged over twenty or thirty years, and more concerned about the next few years at any given moment. As such, it is important to know the details of what would need to be bought for a given policy change. If, for example, the United States has a surplus of tanks in storage that could be restored to good working condition fairly cheaply, investment costs for creating two new divisions might be relatively modest. But the situation could easily be the reverse as well, depending on the circumstances.

Notes

1. Office of Management and Budget, *Historical Tables: Budget of the U.S. Government, Fiscal Year 2009* (Washington, D.C.: Government Printing Office, 2008), p. 137.

2. International Institute for Strategic Studies, *The Military Balance 2008* (Oxfordshire, England: Routledge, 2008), pp. 443–49.

3. Office of Management and Budget, *Historical Tables*, pp. 61, 79, 134–35.

4. To be somewhat more specific, about 90 percent of outlays for personnel take place in the first year and almost all the remainder in the second. About 48 percent of outlays for O&M take place in the first year (averaging across the services), another 40 percent in year two, about 5 percent in year three, and small amounts in years four through six. Procurement dollars are spent more slowly—only about 20 percent in year one, 40 percent in year two, 27 percent in year three, 8 percent in year four, and smaller amounts in the next three years. For research, development, testing, and evaluation, the figures are roughly 45 percent in year one, 42 percent in year two, 7 percent in year three, and small amounts thereafter. See Under Secretary of Defense (Comptroller), *National Defense Budget Estimates for FY 2009* (Washington, D.C.: Department of Defense, March 2007), pp. 54–55.

5. M. Thomas Davis, *Managing Defense After the Cold War* (Washington, D.C.: Center for Strategic and Budgetary Assessments, 1997), pp. 1–3.

6. See Under Secretary of Defense (Comptroller), *National Defense Budget Estimates for FY 2009*, pp. 80, 102.

7. Raymond Hall, David Mosher, and Michael O'Hanlon, *The START Treaty and Beyond* (Washington, D.C.: Congressional Budget Office, 1991),

pp. 62, 135; Office of the Under Secretary of Defense (Comptroller), *National Defense Budget Estimates for FY 2009* (Washington, D.C.: Department of Defense, 2007), p. 43, available at www.defenselink.mil/comptroller/defbudget/fy2009/fy2009_greenbook.pdf [accessed September 2, 2008]; and Stephen I. Schwartz, ed., *Atomic Audit: The Costs and Consequences of U.S. Nuclear Weapons Since 1940* (Washington, D.C.: Brookings, 1998), p. 3.

8. Naturally, more detailed questions require more detailed analysis. For example, when determining compensation levels needed to ensure a given level of recruiting success, comparison between military and civilian pay levels for individuals of comparable age and skill are needed, as is historical data on the typical correlation between a given pay increase or other improvement in compensation and improved recruiting results. Such comparisons need to go beyond broad metrics to examine various types of compensation levels, and various types of occupational specialties. For example, on balance there is no systematic civilian–military pay gap (especially when all of DoD's compensation, including tax advantages and housing and family subsidies are counted, without even including retirement or health benefits since they do not accrue to all). Counting all direct benefits, a typical twenty-two-year-old person with four years experience and an E-4 ranking earns the equivalent of $75,000 (in 2009 dollars) if unmarried, and about $90,000 if married with two children (for the former, for example, 33 percent of compensation is in basic pay, 18 percent in basic allowances for housing and subsistence, 3 percent in tax advantages, 8 percent in health care and other noncash benefits, and a total of 38 percent in deferred benefits [depending on the person], which include veterans' benefits, retiree health care, and retirement pay). However, certain individuals with specific technical skills may well make substantially less in the military than they would in the private sector. See Carla Tighe Murray, *Evaluating Military Compensation* (Washington, D.C.: Congressional Budget Office, 2007), pp. 22, and 31–32; and Richard L. Fernandez, *What Does the Military 'Pay Gap' Mean?"* (Washington, D.C.: Congressional Budget Office, 1999), pp. 1–7.

9. Congressional Budget Office, "Analysis of the Growth in Funding for Operations in Iraq, Afghanistan, and Elsewhere in the War on Terrorism," U.S. Congress, Washington, D.C., February 2008, p. 1, available at www.cbo.gov [accessed September 2, 2008]; Steven M. Kosiak, "The Cost of U.S. Operations in Iraq and Afghanistan and for the War on Terrorism through Fiscal Year 2007 and Beyond," CSBA Update, Center for Strategic and Budgetary Assessments, Washington, D.C., September 12, 2007, pp. 1–5; and Testimony of Steven M. Kosiak before the Senate Budget Committee, "The Global War on Terror (GWOT): Costs, Cost Growth and Estimating Funding Requirements," Center for Strategic and Budgetary Assessments, Washington, D.C., February 6, 2007, pp. 1–8.

10. Kosiak, "The Global War on Terror," pp. 3–6.

11. Congressional Budget Office, "Costs of Operation Desert Shield," CBO Staff Memorandum, Congressional Budget Office, Washington, D.C., January 1991, pp. 1–20; and Department of Defense, *Conduct of the Persian Gulf War: Final Report to Congress* (Washington, D.C.: Department of Defense, 1992), p. P-2.

12. Kosiak, "The Global War on Terror," p. 6.

13. David Newman and Jason Wheelock, "Analysis of the Growth in Funding for Operations in Iraq, Afghanistan, and Elsewhere in the War on Terrorism," Congressional Budget Office, February 11, 2008, pp. 1–3.

14. Robert F. Hale, *Promoting Efficiency in the Department of Defense: Keep Trying, But Be Realistic* (Washington, D.C.: Center for Strategic and Budgetary Assessments, 2002).

15. Elise Castelli, "Study: Intel Contract Employees Costly," *Defense News*, September 1, 2008, p. 24.

16. Matthew Goldberg, *Logistics Support for Deployed Military Forces* (Washington, D.C.: Congressional Budget Office, October 2005), pp. 27–44; and Daniel Frisk and R. Derek Trunkey, *Contractors' Support of U.S. Operations in Iraq* (Washington, D.C.: Congressional Budget Office, August 2008), pp. 1, 8, 13 (of 22), available at www.cbo.gov/ftpdocs/96xx/doc9688/MainText.3.1.shtml [accessed August 20, 2008]. About 190,000 contractors were working in Iraq in early 2008—just under 40,000 Americans, about 70,000 Iraqis, and about 80,000 third-country nationals. In most other wars, the United States had about one contractor for every five troops; in Korea, the ratio was 1 to 2.5.

17. The greatest inefficiencies occur when a production line is sized, scaled, and built for a certain total procurement buy, only to have that buy reduced in numbers thereafter. But there are penalties even when a production line can be redesigned for a smaller procurement lot. For example, according to one estimate based on CBO data and other sources, a hypothetical two-thirds reduction in the Air Force's planned purchase of F-35 Lighting II aircraft (also known as the joint strike fighter), from roughly 1,700 planes to 500, suggested a 15 percent unit price increase. Also, a cut in the F-22 purchase from the intended 340 to about 125 was estimated to increase unit costs some 40 percent. See Michael O'Hanlon, "The Plane Truth," *Brookings Policy Brief* #53 (Washington, D.C.: Brookings, September 1999), p. 6. More generally, according to CBO, a 50 percent reduction in the annual rate of production for aircraft increases production costs by roughly 10 percent to 35 percent (15 to 30 percent for most planes)—depending in part on whether the lower rate of production still remains above the "minimum economic rate" range or drops below it. A 50 percent reduction in the production rate for Army vehicles implied a 25 to 40 percent increase in unit cost, roughly, and for missiles, the typical unit cost increase was 15 to 50 percent. See R. William Thomas, *Effects of Weapons Procurement Stretch-Outs on Costs and Schedules* (Washington, D.C.: Congressional Budget Office, 1987), pp. 17–18.

18. See Frances M. Lussier, *Structuring the Active and Reserve Army for the 21st Century* (Washington, D.C.: Congressional Budget Office, 1997), p. 3. At that time, for example, the Active Army had 176,000 soldiers in combat units, another 136,000 in deployable support units, and 183,000 in institutional and other nondeployable categories. For the National Guard, the respective numbers in the three categories were 175,000, 152,000, and 40,000; for the Army Reserves, they were zero, 139,000, and 69,000, respectively.

19. William W. Kaufmann, *Assessing the Base Force: How Much Is Too Much?* (Washington, D.C.: Brookings, 1992), pp. 70, 74, 82.

20. See, for example, Office of the Under Secretary of Defense (Comptroller), *National Defense Budget Estimates for FY 2008* (Washington, D.C.: March 2007), p. 78.

21. Mark V. Arena, Obaid Younossi, Kevin Brancato, Irv Blickstein, and Clifford A. Grammich, *Why Has the Cost of Fixed-Wing Aircraft Risen?* (Santa Monica, Calif.: RAND, 2008), p. xvii.

22. One basic approach to making this estimate is first to see how much each service's inflation-adjusted budget has grown since the period that Kaufmann studied. Kaufmann's book was done in 1992, but he took projected forces and costs for 1997 as his focus of comparison, so the question is how much did real budget authority increase from 1997 through 2009 (not counting supplemental costs for Iraq and Afghanistan). The answer is about 45 percent for the Army, 30 percent for the Navy/Marines, and nearly 40 percent for the Air Force. Then, changes in force structure must be accounted for; these increased budgets now generally fund *smaller* forces, meaning the unit cost growth is even greater. Army force structure has on balance declined by just over 10 percent relative to planned base force levels, Air Force force structure has been reduced about one third, Marine Corps force structure has held relatively steady, and the Navy (excluding the carrier force) has declined in size by about one-third. See Office of the Under Secretary of Defense (Comptroller), *National Defense Budget Estimates for FY 2008*, pp. 125–27; International Institute for Strategic Studies, *The Military Balance* 2007 (Oxfordshire, England: 2007), pp. 28–38; International Institute for Strategic Studies, *The Military Balance 1997/98* (Oxford: Oxford University Press, 1997), pp. 18–27; and International Institute for Strategic Studies, *The Military Balance 1992–1993* (London: Brassey's, 1992), pp. 18–28.

23. Raymond Hall, David Mosher, and Michael O'Hanlon, *The START Treaty and Beyond* (Washington, D.C.: Congressional Budget Office, 1991), pp. 139–40.

24. Amy Belasco, *Paying for Military Readiness and Upkeep: Trends in Operation and Maintenance Spending* (Washington, D.C.: Congressional Budget Office, September 1997), pp. 5, 14; Allison Percy, *Growth in Medical Spending by the Department of Defense* (Washington, D.C.: Congressional Budget Office, September 2003), p. 2; Gregory T. Kiley, *The Effects of Aging on the Costs*

of Operating and Maintaining Military Equipment (Washington, D.C.: Congressional Budget Office, August 2001), pp. 1–2; and Secretary of Defense Robert Gates, "FY 2008 President's Budget for Defense," Department of Defense, Washington, D.C., February 5, 2007, p. 5.

25. Lane Pierrot and Gregory T. Kiley, *The Long-Term Implications of Current Defense Plans* (Washington, D.C.: Congressional Budget Office, January 2003), pp. 44–46; Rachel Schmidt, *An Analysis of the Administration's Future Years Defense Program for 1995 through 1999* (Washington, D.C.: Congressional Budget Office, January 1995), pp. 41–44; and Adam Talaber, *The Long-Term Implications of Current Defense Plans and Alternatives: Summary Update for Fiscal Year 2006* (Washington, D.C.: Congressional Budget Office, October 2005), pp. 20–22. In the latter report, CBO estimates that total DoD investment costs (development plus procurement) might be about 15 percent more than anticipated, allowing for the risk of growing costs—but this calculation includes all acquisition including ongoing production of systems for which costs are already well understood. For new systems, average cost increases are thus typically 25 percent (or more) in both the development and procurement phases.

26. Lane Pierrot and Jo Ann Vines, *A Look at Tomorrow's Tactical Air Forces* (Washington, D.C.: Congressional Budget Office, January 1997), pp. 84–89.

27. Pierrot and Kiley, *Long-Term Implications of Current Defense Plans*, p. 43.

28. Adam Talaber, *The Long-Term Implications of Current Defense Plans and Alternatives: Summary Update for Fiscal Year 2006* (Washington, D.C.: Congressional Budget Office, October 2005), pp. 1–2.

29. Quoted in Deborah Clay-Mendez, Richard L. Fernandez, and Amy Belasco, *Trends in Selected Indicators of Military Readiness, 1980 Through 1993* (Washington, D.C.: Congressional Budget Office, 1994), p. 1, citing the definition from the Joint Chiefs of Staff, *The Dictionary of Military and Associated Terms*, JCS Publication 1 (Washington, D.C.: Department of Defense, 1986).

30. Richard K. Betts, *Military Readiness: Concepts, Choices, Consequences* (Washington, D.C.: Brookings, 1995), pp. 115–43.

31. Betts, *Military Readiness*, pp. 43–62.

32. Ibid., pp. 87–114.

33. Under Secretary of Defense (Comptroller) Tina W. Jonas, "Summary Justification: Fiscal Year 2009 Budget Request," Department of Defense, Washington, D.C., February 4, 2008, p. 14.

34. Thomas E. Ricks, *Fiasco: The American Military Adventure in Iraq* (New York: Penguin, 2006), pp. 149–202; Susan L. Marquis, *Unconventional Warfare: Rebuilding U.S. Special Operations Forces* (Washington, D.C.: Brookings Institution, 1997), pp. 1–5; Frederick W. Kagan, *Finding the Target: The Transformation of American Military Policy* (New York: Encounter Books, 2006), pp. 92–100; and Ivo H. Daalder and Michael E. O'Hanlon, *Winning Ugly: NATO's War to Save Kosovo* (Washington, D.C.: Brookings, 2000), pp. 125–26.

35. Some wartime operations are less vindications of readiness—again, the performance of tasks and missions that have in some sense been anticipated and planned for—than of real-time military innovation. In Operation Desert Storm in 1991, the U.S. armed forces developed an aerial tactic known as "tank plinking" in which laser-guided bombs were used against stationary Iraqi tanks that had been located on the desert floor by infrared detectors not originally intended for such a purpose. In short, the military learned it was not as ready to attack dug-in tanks as it might have been, but it soon found a way to become more effective. The Afghanistan mission of 2001 required real-time development of new tactics involving special forces, as well as local allies, in conjunction with another air-dominant American campaign plan. See Stephen Biddle, *Military Power: Explaining Victory and Defeat in Modern Battle* (Princeton, N.J.: Princeton University Press, 2004), pp. 199–201; and Michael R. Gordon and General Bernard E. Trainor, *The Generals' War: The Inside Story of the Conflict in the Gulf* (Boston: Little Brown and Company, 1995), pp. 322–23.

36. David Von Drehle, "Cheney Steps Up Criticism of Military Readiness," *The Washington Post*, August 31, 2000, p. A1.

37. Steven Lee Myers, "What War-Ready Means, in Pentagon's Accounting," *The New York Times*, September 4, 2000, p. A11.

38. Clay-Mendez, Fernandez, and Belasco, *Trends in Selected Indicators of Military Readiness*, pp. 9–14.

39. Heidi Golding and Adebayo Adedeji, *Recruiting, Retention, and Future Levels of Military Personnel* (Washington, D.C.: Congressional Budget Office, 2006), p. 6; and John Allen Williams, "Anticipated and Unanticipated Consequences of the Creation of the All-Volunteer Forces," in McCormick Tribune Conference Series, *The U.S. Citizen-Soldier at War: A Retrospective Look and the Road Ahead* (Chicago, Illinois: McCormick Tribune Foundation, 2008), pp. 37–38.

40. Anita Dancs, "Military Recruiting 2007: Army Misses Benchmarks by Greater Margin," National Priorities Project, January 22, 2008, available at www.nationalpriorities.org/militaryrecruiting2007 [accessed April 2, 2008].

41. Lt. Col. Bryan Hilferty, "Information Paper: West Point Graduate Retention After 5-Year Active Duty Service Obligation," West Point, New York, December 5, 2007.

42. U.S. Army Fact Sheet, "U.S. Army Officer Retention Fact Sheet as of May 25, 2007," U.S. Army, Washington, D.C., May 25, 2007, available at www.armyg1.army.mil/docs/public%20affairs/officer%20retention%20fact%20sheet%2025may07.pdf [accessed March 25, 2008].

43. Stephen J. Lofgren, "Retention During the Vietnam War and Today," U.S. Army Center of Military History Information Paper, U.S. Army, Washington, D.C., February 1, 2008.

44. Michele A. Flournoy, "Strengthening the Readiness of the U.S. Military," Testimony before the House Armed Services Committee, February 14, 2008, p. 3.

45. "U.S. Army Officer Retention Fact Sheet as of May 25, 2007," available at www.armyg1.army.mil/docs/public%20affairs/Officer%20Retention%20Fact%20Sheet%2025May07.pdf [accessed April 2, 2008].

46. Heidi Golding and Adebayo Adedeji, *The All-Volunteer Military: Issues and Performance* (Washington, D.C.: Congressional Budget Office, July 2007), pp. 14–17.

47. Ann Scott Tyson, "Military Waivers for Ex-Convicts Increase," *The Washington Post*, April 22, 2008, p. A1.

48. Leslie Kaufman, "After War, Love Can Be a Battlefield," *The New York Times*, April 6, 2008, p. ST1; and Pauline Jelinek, "Military Divorce Rate Holding Steady," WTOPNEWS.com, March 1, 2008, available at www.wtopnews.com [accessed April 1, 2008].

49. Pauline Jelinek, "Army Suicides Highest in 26 Years," Washingtonpost.com, August 15, 2007, available at www.washingtonpost.com/wp-dyn/content/article/2007/08/15/AR2007081502027_pf.htm [accessed April 1, 2008]; "Suicide Statistics," available at www.suicide.org/suicide-statistics.htm/#death-rates [accessed April 15, 2008]; and Associated Press, "U.S. Army: Soldier Suicide Rate May Set Record Again in 2008," *The Examiner*, September 5, 2008, p. 12.

50. Thom Shanker, "Army Is Worried by Rising Stress of Return Tours," *The New York Times*, April 6, 2008, p. A1.

51. Laurinda Zeman, *Making Peace While Staying Ready for War: The Challenges of U.S. Military Participation in Peace Operations* (Washington, D.C.: Congressional Budget Office, 1999), p. xiii.

52. Department of Defense, *Quarterly Readiness Report to the Congress, April–June 2000* (Washington, D.C.: Department of Defense, August 2000).

53. Carla Tighe Murray, *Evaluating Military Compensation* (Washington, D.C.: Congressional Budget Office, 2007), pp. 1–20.

54. Steven M. Kosiak, *Military Compensation: Requirements, Trends and Options* (Washington, D.C.: Center for Strategic and Budgetary Assessments, 2005), pp. 20–21; Cindy Williams, "Introduction," in Cindy Williams, ed., *Filling the Ranks: Transforming the U.S. Military Personnel System* (Cambridge, Mass.: MIT Press, 2004), pp. 16–20; and Paul F. Hogan, "Overview of the Current Personnel and Compensation System," in Williams, ed., *Filling the Ranks*, pp. 49–51.

55. For myself, in terms of policy recommendations, the above leads to three judgments: 1) we waited too long as a nation to increase the size of the Army and Marines after 2003, and we should keep increasing the size of the standing active ground forces today although we are belated in doing so; 2) we do not have the luxury of increasing overall deployed force levels abroad now, relative to where they were in late 2008, and we should be trying to draw

down, though the situation does not require this immediately as a matter of extreme urgency; and 3) we owe it to our men and women in uniform to further improve compensation and other help for them and their families, even if benefits are already relatively robust overall.

56. In 1990, the Air Force budgeted for 19.5 hours per air crew per month of flying, and the Navy for 24. In 2008/2009, the respective figures were down to roughly fourteen and eighteen a month. The decline was gradual; about half occurred during the Clinton years, the other half during the George W. Bush years. See Tamar A. Mehuron and Heather Lewis, "Defense Budget at a Glance," *Air Force Magazine* (April 2008), p. 61.

57. Frances M. Lussier, *Replacing and Repairing Equipment Used in Iraq and Afghanistan: The Army's Reset Program* (Washington, D.C.: Congressional Budget Office, 2007), pp. 1–15.

58. International Institute for Strategic Studies, *The Military Balance 2007* (Oxfordshire, England: Routledge, 2007), pp. 406–11.

59. Office of Management and Budget, *Budget of the U.S. Government, Fiscal Year 2008: Historical Tables* (Washington, D.C.: Government Printing Office, 2007), pp. 133–36.

60. International Institute for Strategic Studies, *The Military Balance 2007*, pp. 357, 406–11; and Office of the Under Secretary of Defense (Comptroller), *National Defense Budget Estimates for FY 2008*, p. 134.

61. 9-11 Commission report, Kathleen Ridolfo, "Iraq: Smuggling, Mismanagement Plaguing Oil Industry," Radio Free Europe/Radio Liberty, Washington, D.C., November 13, 2007, available at www.rferl.org/featurearticle/2007/11/38c235c1-6f71-46ac-9463-4119f5cb6fea.html [accessed March 2, 2008].

62. International Institute for Strategic Studies, *The Military Balance 2007*, pp. 406–11; and Office of the Under Secretary of Defense (Comptroller), *National Defense Budget Estimates for FY 2008*, pp. 214–15.

63. International Institute for Strategic Studies, *The Military Balance 2007* (Oxfordshire, England: Routledge, 2007), pp. 340–41; International Institute for Strategic Studies, *The Military Balance 2008* (Oxfordshire, England: Routledge, 2008), p. 376; and Keith Crane, Roger Cliff, Evan Medeiros, James Mulvenon, and William Overhalt, *Modernizing China's Military: Opportunities and Constraints* (Santa Monica, Calif.: RAND Corporation, 2005), pp. 101–3.

64. Department of Defense, *Military Power of the People's Republic of China, 2007: Annual Report to Congress* (Washington, D.C.: 2007), pp. 25–29, available at www.defenselink.mil/pubs/pdfs/070523-China-Military-Power-final.pdf [accessed June 10, 2007].

65. Department of Defense, *Military Power of the People's Republic of China, 2008: Annual Report to Congress* (Washington, D.C.: 2008), p. 32, available at www.defenselink.mil/pubs/pdfs/China_Military_Report_08.pdf [accessed March 10, 2008].

66. See Eric J. Labs, *Increasing the Mission Capability of the Attack Submarine Force* (Washington, D.C.: Congressional Budget Office, March 2002), p. 41.

67. Ibid., pp. 10–13, 41.

Key References and Suggestions for Further Reading

Betts, Richard K., *Military Readiness: Concepts, Choices, Consequences* (Washington, D.C.: Brookings, 1995).

Crane, Keith, Roger Cliff, Evan Medeiros, James Mulvenon, and William Overhalt, *Modernizing China's Military: Opportunities and Constraints* (Santa Monica, Calif.: RAND Corporation, 2005).

Department of Defense, *Conduct of the Persian Gulf War: Final Report to Congress* (Washington, D.C.: Department of Defense, 1992).

———, *Military Power of the People's Republic of China, 2008: Annual Report to Congress* (Washington, D.C.: 2008), available at www.defenselink.mil/pubs/pdfs/China_Military_Report_08.pdf.

Golding, Heidi, and Adebayo Adedeji, *Recruiting, Retention, and Future Levels of Military Personnel* (Washington, D.C.: Congressional Budget Office, 2006).

International Institute for Strategic Studies, *The Military Balance 2008* (Oxfordshire, England: Routledge, 2008).

Kaufmann, William W., *Assessing the Base Force: How Much Is Too Much?* (Washington, D.C.: Brookings, 1992).

Kosiak, Steven M., *Military Compensation: Requirements, Trends and Options* (Washington, D.C.: Center for Strategic and Budgetary Assessments, 2005).

Labs, Eric J., *Transforming the Navy's Surface Combatant Force* (Washington, D.C.: Congressional Budget Office, 2003).

Lussier, Frances M., *Replacing and Repairing Equipment Used in Iraq and Afghanistan: The Army's Reset Program* (Washington, D.C.: Congressional Budget Office, 2007).

Newman, David, and Jason Wheelock, "Analysis of the Growth in Funding for Operations in Iraq, Afghanistan, and Elsewhere in the War on Terrorism," Congressional Budget Office, February 11, 2008.

Under Secretary of Defense (Comptroller), *National Defense Budget Estimates for FY 2008* (Washington, D.C.: Department of Defense, March 2007).

Williams, Cindy, ed., *Filling the Ranks: Transforming the U.S. Military Personnel System* (Cambridge, Mass.: MIT Press, 2004).

CHAPTER II

Modeling Combat and Sizing Forces

Is IT POSSIBLE to make meaningful estimates of how wars will unfold? The most important result to try to predict is, of course, the winner. Even in situations where that might be rather foreseeable, as in the two U.S.-led invasions of Iraq in the last two decades, gauging the likely duration of the conflicts—and the likely casualties that will ensue to participating armies and proximate civilians—is of great interest when possible.

An important related question is: Can we make meaningful estimates of what force package would be adequate to prevail, and prevail decisively, in a proposed conflict? This question gets to the heart of force planning and defense budget analysis, for the United States and also for other countries. It should also influence decisions on whether to enter into war in the first place, for situations in which there is a real choice of when or whether to fight.

In the abstract, predicting outcomes in war is extremely hard. As Australian historian Geoffrey Blainey notes, for example, "when nations prepare to fight one another, they have contradictory expectations of the likely duration and outcome of the war. . . . it is doubtful whether any war since 1700 was begun with the belief, by *both* sides, that it would be a long war. . . . No wars are unintended or 'accidental.' What is often unintended is the length and bloodiness of the war. Defeat too is unintended."[1]

There is a reason so many people have been so unsuccessful in predicting the course of armed conflict. It is because war depends greatly on variables that are very hard if not impossible to quantify, such as the quality of leadership, the effectiveness of any surprise, and the performance of new weapons systems or military operational concepts not previously tested in battle (and hence not well understood in advance of battle).

According to historical data sets, even if one country or alliance is clearly strongly than another according to various military metrics such as overall manpower or combat equipment inventories or a combination of both, it is very hard to make high-confidence predictions about which side will win. Brookings senior fellow Joshua Epstein compiled evidence from the U.S. Army Concepts Analysis Agency published in the 1980s that made this point clearly. The Army's Concepts Analysis Agency had gathered data on many past wars, and used that data to compute overall scores on the capabilities of attackers and defenders through a process described in more detail in the following. It turned out that for countries with roughly comparable military capabilities, with neither side having more than a 50 percent edge against the other in military inputs, the attacker won 142 out of 230 battles, or 58 percent of the total.[2] There was no tendency towards stalemate (only 7 percent of all cases). Nor was there a major inherent advantage for the defender (who prevailed in just 35 percent of all cases). Benefiting from tactical surprise and innovative concepts about how to fight, the attacker usually won, but no other broad conclusion could be reached from the dataset.[3]

Moreover, even though the attacking country usually won, it was far from assured of doing so. The attacking side won only 63 percent of the time when having an estimated advantage between 50 percent and 200 percent over its opponent, and 74 percent of the time when having an even greater edge.[4] So even when military balances seem to clearly favor one side, outcomes are hard to forecast. This fact helps explain arms race dynamics. It is not easy to measure a military balance accurately. Correlations of forces always depend on specifics of when and where a war is fought. In addition, substantial initial advantages are needed to provide any real confidence about the likely outcome. In such a situation, it is no surprise that rival countries with grievances between them sometimes enter into escalating spirals of military buildups.

The difficulty of measuring balances accurately, and the need for a substantial advantage to provide any confidence of victory, help explain why the United States is so bent on having a substantial military edge in any battle. (As noted in the following, its advantages in recent conflicts have, when quantified, typically been three or four or five to one against likely foes without even counting airpower.) Few seriously doubt the likely outcome for such invasions by an overwhelming power against a far smaller, weaker, and less sophisticated power—at least when the outcome is defined in terms of who "captures the flag" or winds up in control of the capital city and major infrastructure. But that is only because the disparity in capabilities is so great. In such cases of extreme mismatch, it is even possible to hazard estimates about the likely duration and likely casualty levels from battle. Even if predictions are naturally imprecise, they can often predict the order of magnitude correctly (that is, actual results will be within a factor of ten of estimated results, and often within a factor of two or three). It has been recognized for years that expected casualties are generally an important consideration when Americans make decisions about whether and how to go to war, meaning that predictions about such matters can be important on policy grounds.[5]

Of course, as the case of Iraq also underscores, predicting the outcome of classic large-scale military engagements is one thing. Predicting casualties—or even the winner—in counterinsurgency missions and "irregular combat" is something else. So if the business of predicting the results of war is very hard—with errors of 100 to 200 percent amounting to relatively *good* results, and with far larger inaccuracies common when foes are nearly evenly matched or when battle involves guerrilla, terrorist, and urban operations—why bother with the business of modeling combat at all?

One reason is that, in science as well as other disciplines, understanding what we *do not* or *cannot* know is important in its own right. If a large body of human experience strongly suggests that warfare is an inherently risky and unpredictable business, the onus will be on policymakers who propose engaging in warfare to explain why it is necessary—and to explain how they have taken every reasonable precaution to minimize the chances of being surprised by the course of events. A good case in point is the Bush administration's decision to engage in what some have called a "war of choice" to overthrow Saddam in 2003—and even more, its decision to do so with smaller forces than recommended by military officers or other analysts, and without a well-developed plan for keeping the peace and restoring order after Saddam was toppled. Greater care in preparation, and less confidence about the probable course of battle, would have been appropriate.

Of course, modeling does not always produce pessimistic forecasts. But it should at least explain and underline the types of assumptions needed to make a certain prediction. In preparing their devastating 1967 surprise attack on Arab air forces, Israeli planners realized that they might be able to achieve a remarkable success, based on a sense of how accurately bombs could be dropped, how easily enemy radar could be evaded, and how vulnerable airplanes and airfields would be to ordnance of a given size and explosive power. But they also had to appreciate how many things would have to go right, in terms of the timing and execution of the attack, to make it succeed. That helped focus them on making preparations very carefully. To take another example, when my former colleague at the Congressional Budget Office, Lane Pierrot, presciently argued in 1990 that it was possible U.S. aircraft loss rates in Operation Desert Storm would be as low as peacetime training rates, she was making the sophisticated yet simple argument that American countermeasures and flying practices might be good enough to counter much of the Iraqi air defense network. Pierrot foresaw that American airpower might render the latter air defenses largely powerless except when coalition aircraft had to fly low to go after certain targets, or penetrate particularly dense air defense bastions. But Pierrot also realized that losses could be much higher; she wound up accurately prognosticating the results in part because she did not seek to be *too* accurate, but focused instead on establishing plausible lower and upper bounds for expected losses. This is an important lesson about the proper way to model combat.

A second reason to model is that we actually cannot avoid it even if we wish to. As Joshua Epstein has convincingly argued, whether we "model" mathematically and systematically, or anecdotally and impressionistically, *everyone* who forms an opinion on whether a given war should be fought is in effect predicting its outcome or at least its plausible range of outcomes. In other words, everyone is operating from some image, some set of expectations, of the likely course of battle, whether precise or not. Otherwise, there would be no way to decide if a given war should be fought in the first place, or if 100,000 or 500,000 troops might be needed for it, or if the nation's industry should be directed to prepare for a long struggle and if a general mobilization might be required.[6] The issue is not really whether we try to find a simplified construct for predicting battle outcomes—all of us do; in fact, all of us must. The issue is whether we choose to employ impressionistic and purely subjective "modeling" or a more rigorous and formal approach. The advantage of formal modeling is that it requires one to make assumptions explicit, and justify them as well as possible using historical, technical, and operational data.

Elaborate computer simulations are not always the best way to model combat. To be sure, algorithms like the Institute for Defense Analyses' TACWAR are always worth consulting. They explicitly simulate the use of multiple weapons over complex terrain, in variable weather conditions, and with numerous possible assumptions about the performance of airpower, logistics systems, and other key determinants of battlefield outcomes. Even these types of models generally have limits on their ability to mimic reality, however. TACWAR has in the past only allowed for a simple front line where armies encounter each other, rather than a dynamic and complex battlefield, and has limited the user's ability to make various assumptions about the possible performance of different weapons.[7] For these and other reasons, the complex models do not always produce the best predictions. Simpler calculations about the likely course of Operation Desert Storm in 1991 done by numerous outside analysts were generally more accurate than sophisticated Pentagon computer runs. While virtually all were too pessimistic, outside analysts generally estimated that U.S. fatalities would total around 1,000 whereas more elaborate models reportedly projected American deaths up to several times as numerous. (Actual combat losses were 146, and including all phases of the operation nearly 400 Americans lost their lives.)[8]

This chapter employs several different simple frameworks for modeling combat. All involve some amount of basic arithmetic. In no case should the use of math distract the reader from the need to critically evaluate assumptions, to think hard about where calculations may go wrong, and to remember how often warfare surprises us.

The discussion begins with fairly classic and simple models of traditional combat, starting with Lanchester's equations. It then considers other models, some embodied in simple formulas and others best understood as a systematic way of thinking through a problem or scenario. In other words, the mathematics are not always paramount, and they are not always neatly condensed into one or two simple algorithms. After Lanchester, slightly more complicated models for armored combat are discussed. Then, other types of ground combat—infantry war, urban war, counterinsurgency—are considered together. The focus then changes to naval combat, including a scenario for amphibious assault and another for a blockade operation at sea. The old-fashioned (and hopefully obsolete) matter of nuclear exchange calculations is then discussed. The chapter concludes with a framework for analyzing progress in counterinsurgency operations like those in Iraq and Afghanistan.

Lanchester's Equations

Although they are somewhat simplistic, the Lanchester equations are a good place to start in understanding how war simulations are conducted and combat models created. Formulated a century ago by a British engineer who gave the equations their name, they have the advantage of simplicity—since war was, on balance, somewhat simpler then, if still inherently hard to predict.

Lanchester equations come in several forms, but two are of particular note: those for direct fire (like rifles aimed at specific individuals) and indirect fire (such as artillery fired into a broad area where enemy forces are known to be located even if they are not individually visible and targetable).

Of course, modern war has elements of both types of fire, so any sophisticated model would need to combine their effects (accounting as well for a variety of types of weapons, for complex terrain, for command and control, for maneuvers, and so on). But some of the basic dynamics are easier to see if combat is first simplified and abstracted into these two simple forms.

Direct-Fire Equation

Imagine two rows of eighteenth-century soldiers lined up firing at each other. Assume they are not quite shoulder to shoulder, but instead have some spacing between them (making it very unlikely that one soldier will be hurt by a weapon fired at another soldier). Assume further that there is no issue of concealment; all soldiers on both sides are within sight, and weapons range, of each other.

This model of combat may be a relatively good way to understand naval gunfire exchange at sea (more on that appears in a subsequent part of this chapter), certain types of aerial combat in the skies, and in general warfare where there is little subtlety or complexity to the battlefield—enemy forces see each other and try to destroy each other in a fairly straightforward exchange of gunfire.

With this image of combat in mind, and two armies represented by $A(t)$ and $B(t)$, the basic mathematics of the Lanchester equation are simple to derive. (They involve a small amount of very basic calculus; those unable to follow it need not worry since the result can be understood just through arithmetic.) $A(t)$ and $B(t)$ are the total numbers of soldiers on each side still unwounded and capable of fighting at a given time. Clearly, $A(t)$ will diminish faster the larger the number of enemy soldiers on the other side and the more effective that side's weapons. Weapons effectiveness, which can be simplified as the multiplicative product of the rate of fire, the accuracy of the weapon, and the lethality of the weapon, is condensed into a single term for each side (represented by a and b, respectively, for army A and army B). In mathematical form, assuming a simple linear relationship, this can be written as follows (those uninterested in the derivation or put off by the calculus can simply skip a few lines to the actual formula):

$$dA(t)/dt = -bB(t)$$

Similarly, for $B(t)$, we have:

$$dB(t)/dt = -aA(t)$$

In other words, A's forces decline faster to the extent that B has a larger army firing more lethal weaponry at them, and vice versa.

By the chain rule, dA(t)/dt can also be written as:

$$dA(t)/dt = [dA(t)/dB(t)][dB(t)/dt]$$

Substituting the term on the right into the first equation, we get:

$$[dA(t)/dB(t)][dB(t)/dt] = -bB(t)$$

Further substituting for dB(t)/dt from the second equation in this list yields:

$$[dA(t)/dB(t)][-aA(t)] = -bB(t)$$

Rearranging terms and canceling out the respective minus signs we get:

$$a[A(t)][dA(t)] = b[B(t)][dB(t)]$$

Integrating, and expressing for the time t = T, we get:

$$A^2(T) - A^2(\text{initial}) = [b/a][B^2(T) - B^2(\text{initial})]$$

In other words, the square or second power of A(t) is on the left hand side, for two different times, the initial time at the beginning of the battle, and a time T later in the course of events. The right side of the equation has a similar term for B, times the ratio of the typical weapons effectiveness for side B to that for side A.

As one example, assume army A starts with ten soldiers and army B with twenty (at time t=0), and assume weapons effectiveness to be the same on both sides. How many soldiers will army B still have standing when A's force has been wiped out (at time t=T)? This boils down to solving the following arithmetic problem:

$$0^2 - 10^2 = B^2(T) - 20^2$$

10^2 is of course 100 and 20^2 is 400, so this becomes:

$$-100 = B^2(T) - 400$$
$$300 = B^2(T)$$
$$B(T) = \sqrt{300} = 17 \text{ (roughly)}$$

In other words, B loses just three soldiers in annihilating all ten of A's soldiers. The Lanchester direct fire equation gives a great premium to numerical superiority, all other things being equal. (Every soldier in B's army on average faces only half the fire of one soldier on the other side, while every soldier in A's army by contrast is being shot at on average by two soldiers from B, explaining why A is at such a disadvantage in this engagement.)

Indirect Fire

Of course, at least two things are unrealistic about the scenario behind the direct-fire Lanchester equation. First, at least for all eras since the eighteenth century, enemy forces will take cover to get out of the way of weapons. Second, many weapons are fired in a way to barrage an area rather than directly strike a specific individual or vehicle. Pure musket-fire exchanges no longer tend to occur in warfare.

The indirect-fire version of the Lanchester equations corrects for these problems. It does so at the price of going to the other extreme and eliminating any role for direct-fire weaponry. But it is still instructive to see how the math changes, and how the predicted battlefield results can also therefore change. If nothing else, this helps us appreciate the stark differences between battlefield dynamics for wars dominated by direct fire and those dominated by indirect fire. (Capturing both direct and indirect fire in a single Lanchester equation is more realistic, but also much more complicated mathematically. In fact, when trying to capture such complex dynamics, I do not employ Lanchester equations here, but other types, as discussed further on in this chapter.)

In Lanchester's indirect fire equations, the likelihood that one's forces will suffer losses becomes proportionate to three terms. Two are as before, the effectiveness of the enemy's weapons and the number of enemy forces. The third term is the density of one's *own* forces on the battlefield. Having more people means having more targets, increasing the chances that an enemy weapon fired more or less randomly at a broad area will, by chance, strike soldiers in that area upon detonation. So we have:

$$dA(t)/dt = -bB(t)A(t)$$

and

$$dB(t)/dt = -aA(t)B(t)$$

Again, using the chain rule employed earlier:

$$dA(t)/dt = [dA(t)/dB(t)][dB(t)/dt] = -bB(t)A(t)$$

And then substituting the second equation for dB(t)/dt gives:

$$[dA(t)/dB(t)][-aA(t)B(t)] = -bB(t)A(t)$$

Rearranging and simplifying gives:

$$a[dA(t)] = b[dB(t)]$$
$$A(T) - A(\text{initial}) = [b/a][B(T) - B(\text{initial})]$$

With this formulation, there is much less benefit to numerical superiority, as having more troops on the battlefield gives one more shooters but also provides the enemy with more targets. As such, the effectiveness of weaponry becomes just as important mathematically as a quantitative advantage in soldiers, something that was *not* true in the direct-fire equation. Mathematically, the number of one's own forces is taken only to the first power, and not squared, as a result.

As one simple example, if A and B have equally effective weapons, and A starts with ten soldiers while B begins with twenty, they will lose personnel at the same rate. Army A will have five people left when B has fifteen left; A will run out of soldiers when B has ten still standing. Again, B wins, but less dramatically. And in a situation where A has somewhat better weapons than B, say 2.1 times as good, it could win. By contrast, in the direct-fire equations, it would need to possess a huge (fourfold) advantage in weaponry in order to compensate for its fewer forces and prevail in the engagement.

Predicting Air-Ground Combat

Simple models for predicting the course of heavy armored combat can best be explained with reference to a specific example such as Iraq. That is because a couple of simple formulas tend to be used together in these, and it is easiest to see how the overall calculation is done with reference to a concrete example rather than abstractions.

Prior to Operation Desert Storm in 1991, a number of scholars, using models and databases developed largely for assessing the NATO–Warsaw Pact military balance during the Cold War, estimated the losses likely to result in a war to expel Iraqi forces from Kuwait. Virtually all these estimates were too high, but they were also generally more accurate than those produced by the Pentagon before the U.S.-led war against Iraq began. Indeed, they were virtually all correct in predicting a short decisive conflict in which U.S. casualties would be far less than American losses in the Vietnam or Korean wars. In that sense, the flawed estimates were still useful.[9]

The Kugler–Posen and Epstein Models

At least two main families of models were developed during this time period in the open literature. As I would define and group them, they might be termed the Kugler–Posen "attrition-FEBA expansion" model and the Epstein adaptive dynamic model.[10] Both are more sophisticated than the famous century-old Lanchester equations, which as noted require simplifying assumptions about the nature of weaponry that apply much better to eighteenth-century musket fire, nineteenth-century battleship duels, or World War I artillery exchanges than to the modern battlefield.[11] They are much less sophisticated than the detailed, and classified, computer models such as "TACWAR" and "Janus" used by the Pentagon community to predict combat outcomes. But they do offer simplicity and accessibility.[12] And again as noted earlier, they also have every bit as good a track record in recent times of predicting combat outcomes. What they may lack in detail and exactitude they tend to make up for by requiring a user to think pragmatically, historically, and intuitively about the modeling enterprise—rather than running the risk of getting lost in the math, or being overly impressed by the internal machinations of the complex computer programs.

The first of these unclassified and relatively simple models, developed by Richard Kugler (of National Defense University in Washington) and Barry Posen (of MIT), was optimized for a war of attrition in which NATO was presumed to be on the defensive. It is based on the assumption that a military of sufficient size can hold a front of a given length against all-out enemy assault. Provided that the defender can reinforce its losses and maintain an adequate "force-to-space ratio" as the forward edge of the battle area (or FEBA) evolves, it should be able to hold the line and protect its territory.[13] Posen later applied the model to the U.S. plan to liberate Kuwait from Iraq, based on the assumption that such a war would resemble a NATO–Warsaw Pact confrontation in the types of weaponry and tactics employed. In that case, he was NOT assuming that the Iraqis could hold their positions robustly, but used the equations primarily to calculate relative casualty levels. There was no inconsistency in this approach, since Iraqi forces were unable to reinforce well during an attack, meaning they could be worn down quickly in an attempted breakthrough sector.

The second model, developed by Joshua Epstein at the Brookings Institution, has numerous similarities with the Kugler–Posen framework, but it challenges the idea that a sufficient "force-to-space" ratio ensures a viable defense. That ratio is typically cited as one armored division equivalent (ADE) per twenty-five kilometers of front according to proponents of that approach, though some estimates suggest an ADE can cover twice as much frontage—underscoring the imprecision of this rule of thumb.[14] Epstein also rejects the popular idea that a defense can fend off an attack provided that the offense does not achieve a 3:1 force advantage in the sector of attempted breakthrough.[15] In fact, as he convincingly shows, and as databases like those of the late Trevor Dupuy and the Army's Concepts Analysis Agency demonstrate as well, attackers have often succeeded historically when simply equal to (or even smaller than) the defense.[16]

Epstein's model also allows for the possibility that a defender might withdraw in order to buy time, improve its position, or slow the pace of battle—thereby slowing the rate at which casualties are incurred. In other words, geography and the movement of forces are still part of the analysis, but in a different way, tied to casualty rates rather than force-to-space ratios.[17] The differences in these methods can be quite significant in a given case.

For a war in which breakthrough operations are successful, and in which organized withdrawal is not a major factor, the Posen–Kugler and Epstein approaches can be collapsed into a single approach. This combined approach at least gives a sense of relative casualties on the two sides. That is the method followed here.[18]

In fact, the methods have many commonalities. Both focus on armored divisional equivalents and their heavy weaponry as the main variables in their associated equations. "Armored division equivalents" not only reflect the quantity of armored formations, but also their equipments' quality. They are defined such that a modern U.S. heavy division equals 1.0 ADEs.

The models also specifically incorporate a role for ground-attack aircraft in the close-air support role. They are assumed to be capable of dropping a given number of munitions per sortie, and flying a given number of sorties per day (with a given probability of being shot down on each mission). Each munition is assigned a "kill probability" to reflect the likelihood it will strike and destroy a major ground vehicle. Knowing the number of such ground vehicles per division, one can translate the attrition from aerial attacks into armored division equivalents, thereby linking the ground and air wars conceptually and mathematically.

In modern war, of course, airpower is used against many targets besides vehicles in the field. That was not only true in the massive city bombing campaigns of World War II, and the "carpet bombing" of Vietnam, but also in Operation Desert Storm, the Kosovo war, and other cases.[19]

In Desert Storm, for example, while just over half of all strike missions were against fielded Iraqi forces, the remaining 43 percent focused on airfields, SCUD missile launchers, surface-to-air missile sites and other air defense infrastructure, strategic lines of communication including bridges and port facilities and train lines, military industry, suspected weapons of mass destruction sites, telecommunications, electricity grids, oil assets, and leadership targets. A total of about 1,000 U.S. attack aircraft (plus another 250 or so from allies such as Saudi Arabia and Britain) dropped around 200,000 munitions, just under 10 percent of them precision-guided.[20] In NATO's 1999 air war against Serbia, Operation Allied Force, roughly 7,600 strike missions were flown against fixed targets and 3,400 against mobile targets. A total of about 1,000 NATO aircraft were ultimately employed in the conflict (including strike planes, air superiority fighters, and support aircraft), dropping some 28,000 munitions, of which about 7,000 were precision-guided. Only two fixed-wing aircraft were lost over Serbia, another malfunctioned and crashed in the Adriatic, and two Apache helicopters crashed over Albania, causing America's only two fatalities of the war.[21]

In the models in question, however, such strategic bombing and "battlefield air interdiction" is viewed as preparatory or exogenous; airpower enters into the mathematics only through its effects on ground power (and possibly through its effects on either side's ability to reinforce key positions).

Each of the two models assumes that the attacker drives the pace of battle and that it is able to adjust the pace of combat up to a certain reasonable maximum. That maximum rate of attacker losses is usually 1 percent to 5 percent per day for a division-sized force, with the lower numbers more typical of protracted fighting and the higher end of the range occasionally attained in intensive combat. This is a historically based figure, consistent with most of the experiences of World War II battles, subsequent Arab–Israeli wars, and other conflicts.[22] A user of either model begins by specifying an assumed daily loss rate for the attacker in the ground battle within this range. Epstein's model allows the defender a say in the intensity of battle, too, by allowing the defender to stage a withdrawal that, as noted, reduces the assumed casualty rates on both sides.

Both models then require the user to estimate an "exchange rate." This is a proportionality factor linking the losses of the attacker to those of the defender. This exchange rate reflects first and foremost the quality of troops and equipment on each side of the war. Losses are expressed in terms of armored divisional equivalents (or ADEs), the coin of the realm in these models. Human casualties can then be inferred once ADE losses have been estimated. (Depending on which country's military is at issue, a division usually has 10,000 to 18,000 soldiers, and usually one to three actual divisions are in any given armored divisional equivalent. This is because less sophisticated militaries often need to put two or three of their divisions on the battlefield to create the effectiveness of just one modern Western division, the standard by which an ADE is defined.)

Applying Models to Operation Desert Storm

How did these various models do in estimating the outcome of Operation Desert Storm? It is worth working through the numbers for an important example like this for three reasons. First, doing so shows how the different pieces of the model (notably the ground war component and the air war component) fit together methodologically. Second, doing so provides useful practice with a concrete case. Third, doing so helps one see the likely limits on accuracy—even for a relatively successful case of modeling—but also its potential, assuming that expectations about what modeling can deliver are kept in check.

Both models were used to predict rapid decisive victory by coalition forces, with considerably higher casualties on the Iraqi side than the American side, and in that regard Posen and Epstein were both correct. More specifically, Posen forecast weeks of combat and 4,000 to 11,000 coalition casualties to liberate Kuwait (including dead and wounded).[23] Epstein predicted weeks of combat as well, and a slightly broader casualty range of 3,000 to 16,000 (again, dead and wounded combined).[24] Both made calculations based on the premise of attrition warfare (albeit short-lived attrition warfare), after relatively short air campaigns, given what was known about likely Pentagon war plans at the time. The assumption of attrition warfare was not altogether incorrect, at least in the opening hours of combat, for U.S. Marines and associated forces who penetrated Iraqi defenses and drove towards Kuwait City. It was incorrect for the forces led by the Army's Seventh Corps, which executed the famous "left hook" to the west of Iraqi defenses, outflanking Iraqi forces in their initial movements, though engaging in occasional combat with Saddam's military thereafter.[25]

Press reports suggested that the Pentagon was prepared for 30,000 or more casualties in Operation Desert Storm, even though it presumably did have access to detailed battle plans when making its predictions.[26] For such estimates, one would presume that 15 to 20 percent of all U.S. casualties would have resulted in deaths and the rest in wounded personnel, given historical trends (that account for the improving benefits of modern medical care).[27]

In the actual event, losses were less than forecast. By official count, a total of 382 Americans died in the southwest Asian theater in the course of Operation Desert Shield, which began in August 1990, and Desert Storm, as that operation was renamed in January 1991. That count includes prewar and postwar accidents and other non-hostile acts. A total of 147 U.S. troops died in combat. Thirty-five were killed accidentally by so-called friendly fire. Others died in accidents of various types, on and off the immediate battlefield. About 500 additional Americans were wounded.[28] Considering allied forces as well, and using round figures, the coalition suffered about 240 combat deaths, some 500 deaths over the course of the entire operation from all causes, and about 1,500 casualties including killed and wounded.[29]

How good were the prewar estimates, and what do the inaccuracies tell us about the value of trying to predict casualties? On the whole, these casualty estimation efforts were rather successful despite their inaccuracies, especially for the broad policy points they implied—that war would be decisive, victorious, and not very bloody by the standards of past major conflicts (yet hardly casualty-free).

With a clearer indication of how long the air war would last prior to the ground campaign, the preceding models could have done an even better job of estimating likely casualties in Desert Storm. And after the fact, it is also possible to adjust various parameters to account for the better-than-expected American performance and worse-than-expected Iraqi performance—providing more accurate tools for modeling any subsequent similar conflict (such as the invasion phase of the 2003 war).

On the latter point, Stephen Biddle has enumerated many of the basic mistakes the Iraqis made. Among other things, they failed to post advance guards near trench lines and failed to remove dirt from the vicinity of those trench lines to keep the locations of dug-in forces secret.[30] While the United States had reconnaissance technologies, such as the Joint Surveillance and Target Attack Radar System (JSTARS), that made it much easier to detect moving Iraqi vehicles, and while infrared detectors helped it find vehicles at certain hours of the evening in particular, it was often unable to know where Iraqi units were through theater-wide surveillance.[31] As such, Iraqis often could have gotten in the first shot, if they had properly exploited their defensive advantages.

In addition, American forces benefited from their supporting superstructure—intelligence, communications, equipment maintenance, and logistics support—even more than expected. The models, focused as they are on individual armored combat units and their traditional weaponry, and dependent on past combat data to generate battlefield performance parameters, do not tend to highlight such capabilities. These facets of modern war give an even greater benefit to a military like the U.S. armed forces, capable as it is of competently assimilating them all into the way it fights, and confer an even greater disadvantage on a country like Iraq that fails to understand or counter them.[32]

High technology, particularly the ability of U.S. airpower to prepare the battlefield for more than a month before the ground war, also played an unanticipated role. For example, the tactic of "tank plinking," in which laser-guided bombs were dropped on Iraqi armor (often in the early evening, when the desert sands cooled more quickly than Iraqi armor, revealing the locations of the latter to infrared sensors), was only developed in the course of the war. It could not have been easily foreseen in a combat prediction done before the war began. The ability of coalition aircraft to undertake that and other effective tactics from high altitude, out of range of Iraq's man-portable surface-to-air missiles, was also not foreseen—even by war planners, who had coalition pilots fly low for the first days of battle. More generally, American military equipment turned out to be even better than expected, compared to Soviet weaponry like that fielded by the Iraqis.[33] As a result of all these factors, the American-led victory over Iraq was far more decisive than Israeli victories in previous wars against Syria, Jordan, and Egypt.[34]

These factors can be adjusted to make future predictions more accurate. The ability of coalition forces to wage an air war indefinitely prior to any ground assault can be reflected in how the models are used, as can the high lethality of modern air-to-ground ordnance.[35] Superior American fighting capability and poor Iraqi competence can be reflected in a lopsided "combat exchange ratio" that further amplifies the adjustments already made due to varying equipment quality. The U.S. ability to stay out of range of much Iraqi fire, at least on the open battlefield, can be reflected in a much lower daily attrition rate for the attacker than usually assumed.[36]

One can get the gist of this without wading through complex calculations. Considerable uncertainty still surrounds the issue of Iraqi losses in Operation Desert Storm. But Iraq appears to have lost roughly 1,100 to 1,400 tanks, about 800 armored personnel carriers, and 1,000 to 1,500 artillery tubes during the air war. That makes about 3,300 major pieces of weaponry. (In a standard U.S. division in the modern era, there have been about 1,200 such major weapon vehicles per division.)[37] It lost another 1,000 to 1,200 tanks, about 700 armored personnel carriers, and 1,000 or more artillery tubes during the ground campaign. These losses came out of initial Iraqi assets of up to 4,000 tanks, 3,000 artillery tubes, and 3,000 armored personnel carriers in the Kuwaiti theater (as well as about 340,000 personnel).[38] Iraqi personnel casualties are even more uncertain, but probably numbered in the low tens of thousands. (Somewhat more than 2,000 Iraqi civilians are also believed to have died in the course of the conflict.)[39]

Knowing this information, we can redo the math to "predict" what happened in Operation Desert Storm more accurately. For the air war, equipment losses resulted from a total use of about 10,000 precision-guided air-to-ground munitions (PGMs) including Maverick and Walleye air-to-surface missiles and laser-guided bombs, as well as from ground fire.[40] Mathematically, the air war is very simple to represent. Assuming a kill probability of about 65 percent per weapon launched (modern norms to that point were typically perhaps 5 to 10 percent),[41] we get:

$$[10{,}000 \text{ (weapons)}][0.65 \text{ (kill probability)}] = 6{,}500 \text{ destroyed vehicles}$$

Of that total of 6,500 vehicles, assume for simplicity that half were main combat vehicles and the other half support vehicles like trucks (a reasonable assumption for most militaries).[42] This translates into about 3,300 Iraqi combat vehicles "predicted" to be lost.

Following the Kugler–Posen and Epstein frameworks, one can then proceed as follows for understanding the ground war. Coalition forces had the equivalent of roughly ten armored divisions in place prior to the outbreak of hostilities (that is, ten ADEs).

By contrast, Iraq's forces included about 2.8 equivalent divisions after the effects of the air war, and once the effects of poor Iraqi technology are factored in.[43] That number is calculated by taking the total number of major Iraqi weapons vehicles, which equaled about 10,000 before the air war, subtracting out the 3,300 vehicles destroyed during the air war (leaving some 6,700 in place), dividing that figure in turn by 1,200 armored vehicles per division (resulting in an estimate of 5.6 total divisions), and then adjusting the 5.6 divisions to account for the lower quality of Iraq's divisions. That last step means dividing by two in this case, resulting in an Iraqi armored "score" of roughly 2.8 ADEs.[44]

For the ground war, one now performs an iterative calculation day by day. Assume that coalition ground forces would lose 0.1 percent of their strength per day, a relatively low total by historical norms for intense combat.

Assume further that they would benefit from roughly a 30:1 combat exchange ratio in terms of "armored division equivalents." It is this step in particular that benefits from hindsight; prior to Desert Storm, the best proxy for a U.S.–Iraqi conflict seemed to be the various Arab–Israeli wars in which Israel often destroyed three to five times as much Arab equipment as it lost from its own military (whether on offense or defense).[45] The fact that the United States could do much better than Israel had done was quite surprising, and helps explain why most modelers overestimated American losses.[46]

On day one, coalition forces would thus lose 0.01 ADEs (0.1 percent of ten ADEs). Iraq would lose thirty times as much, due to the 30:1 combat exchange ratio working against it, or 0.3 ADEs from ground fighting. That is:

$$\text{Iraqi Losses} = [\text{U.S. Losses}][\text{Exchange Ratio}]$$
$$= [0.01 \text{ ADE}][30] = 0.3 \text{ ADEs}$$

On day two, the losses would round to the same amount as for day one (they would actually be slightly less, but only 0.1 percent less).

On day three, again losses would round to the same totals, and the same would be true for day four (this is an unusual result due to the very low coalition loss rate).

In addition, there would be losses from the ongoing air war. If the pace of aerial attack remained roughly as before, the United States might be expected to use another 1,000 munitions during this phase of conflict, destroying 650 Iraqi vehicles in the process. That would translate into 0.27 division equivalents or 0.14 ADEs over the three days in question. (U.S. aircraft losses, while not insignificant, were very low and strategically inconsequential, given the ability of the American armed forces to replenish losses and the limited duration of the war. A total of thirty-eight coalition aircraft were lost and forty-eight damaged by enemy action, almost all from ground-based air defenses—with heat-seeking or infrared surface-to-air missiles as well as anti-aircraft artillery causing more than two-thirds of all attrition.)[47]

So after four days, the approximate duration of the ground campaign in Operation Desert Storm, cumulative losses from the ground war would be 0.04 ADEs for the coalition and 1.34 ADEs for Iraq—meaning about 2.7 Iraqi divisions overall, or 3,200 pieces of heavy weaponry (and perhaps 6,500 vehicles including trucks and the like).

Assuming 18,000 soldiers per ADE, the U.S. losses translate roughly into some 700 casualties, of which about 100 to 150 might be expected to be fatalities and the rest wounded, based on past norms.

In the general case, each day of war could be more complex to model than indicated here. Reinforcements could arrive from outside of the theater, for example. Forces could move on and off the battlefield. Both sides, not just the United States, could use airpower. But for this particular conflict, the preceding math is fairly accurate.

It was reasonable to think that the results of Desert Storm, incorporated into the previous models, would allow relatively accurate predictions of the outcome in 2003. Indeed, they did—for the invasion phase of the war, that is. Most weapons used against armor—laser-guided bombs, Mavericks, and so on—were more or less unchanged relative to 1991.[48] By 2003, the United States had the all-weather satellite-guided joint direct attack munition (JDAM). But the GPS-guided JDAM weapon typically misses its target by five to ten meters, so it is not always sufficiently precise to strike armor.[49] Moreover, weather was not a severe handicap in Desert Storm, so adding all-weather capability to the U.S. PGM inventory might have been expected to make only a marginal difference under similar circumstances in the future.[50] As for the Iraqis, even if they corrected some of the mistakes they made in 1991, they could have been expected to make other mistakes.[51] Moreover, U.S. forces could change tactics in the event that Iraq found a way to hold its own in a given type of firefight, fighting more at night or relying more heavily on attack helicopters or working harder to avoid Iraqi defensive positions rather than driving straight through them. Of course, this logic broke down when another type of warfare, insurgency combined with terrorism, was adopted by the Iraqi resistance.

Modeling Urban and Infantry Warfare

Urban and infantry warfare are often even more challenging than heavy air-ground combat for several reasons. Enemy forces are interspersed among civilian populations that need to be spared as much as possible, on moral as well as strategic grounds, and complex terrain complicates matters as well.

For such combat, a modified and simplified version of the model of the late U.S. Army Colonel Trevor Dupuy can be useful. The Dupuy method does not include a specific means for incorporating the effects of airpower and geography, so in that sense it is less sophisticated than the Kugler–Posen and Epstein models. Its advantage is that it focuses on soldiers, not armored divisional equivalents, making it more useful for infantry combat in which armored formations are generally less central. It is to first approximation an infantry model rather than a heavy air-ground warfare model. As with the Kugler–Posen and Epstein models, it allows the user to modify input data to reflect the quality of each side's troops and equipment. It is also informed by a very detailed dataset on past conflicts. In addition, it incorporates coefficients for a wide range of factors such as weather, surprise, and terrain that require subjective interpretation to employ, but that allow for more explicit consideration of these elements of combat than the other two models.[52]

Dupuy's methodology is a bit hard to follow in its detail, but sensible and logical in its main framework. I simplify it here somewhat. He begins by translating the number of troops fielded by each side into a total power figure, P. P is the product of the size of the fielded force with its quality. It is further modified to account for the degree of surprise achieved in the early days of battle (for an attacker) and to account for the benefits of any concealment, complex terrain, and prepared positions (for a defender).

Using these power figures to calculate relative casualties requires the use of detailed and lengthy tables that reflect Dupuy's experiences with a wide body of combat data from many past battles. In essence, each side's daily casualties are estimated to be the product of three main types of terms: total troop strength, multiplied by a daily maximum casualty rate, multiplied by a factor accounting for the power differentials between the two sides.[53]

In mathematical form, the terms for the attacker and defender are:

$$P_{ATTACKER} = (N_{ATTACKER})(Q_{ATTACKER})(S_{ATTACKER})$$

$$P_{DEFENDER} = (N_{DEFENDER})(Q_{DEFENDER})(S_{DEFENDER})$$

For the attacker, the first term is the number of its troops, the second term is the relative quality of its troops, and the third term represents situational factors of direct relevance to the nature of the attack (notably the ability to achieve surprise). For the defender, the quality term can be set equal to 1 for simplicity (meaning that the attacker term for quality will reflect the disparity of the two sides' capabilities). As noted, the situational terms for the defender include the effects of terrain, prepared positions, and weather. Typically, the situational factor would range from 1 to 2, depending on the degree to which the attacker is aided by surprise or the defender by weather, terrain, and prepared defensive positions. (Each of the individual terms—for weather, for terrain, for the nature of prepared positions, can typically vary by up to 50 or 60 percent, meaning that its representation in a power formula might vary from 1 to 1.6 on average.)

Dupuy's books provide detailed estimates of what the various factors might be for different types of defensive positions, varying degrees of poor weather, and the extent to which the attacker achieves surprise.[54]

The terms for daily casualties might then be (taking 1 percent losses as a "norm" and calculating variance around that norm for each side):

$$C_{ATTACKER} = (0.01)(N_{ATTACKER})(P_{DEFENDER}/P_{ATTACKER})$$

$$C_{DEFENDER} = (0.01)(N_{DEFENDER})(P_{ATTACKER}/P_{DEFENDER})$$

The simplicity of this equation (like all others considered here) is both its strength and its weakness. There is some arbitrariness in the calculation of each side's power, but consulting Dupuy's books gives some historical perspective as to how the qualitative factors can be reasonably estimated. Certainly Dupuy's own personal track record was rather good in using historical analogy and instinct to estimate such factors. It can be harder for someone else with less experience trying to use his approach, but the extensive databases in his books make the process somewhat tractable.

The Dupuy method, with its focus on foot soldiers, seems best suited to infantry battle. Its lack of focus on both airpower and maneuver warfare would suggest it is less useful for modeling heavy combat. It is only fair to note, however, that Dupuy also applied his model to predicting Desert Storm casualties, with accuracy comparable to that of the other two models discussed before.[55] Also, for mixed cases involving heavy combat as well as infantry battle, it can be a useful tool. So it is applied in the following to two cases: the U.S. invasion of Panama in 1989, and the U.S. invasion of Iraq and subsequent effort to stabilize that country beginning in 2003.

Panama

In December 1989, U.S. forces overthrew Panamanian strongman Manuel Noriega and defeated his armed forces. About 22,500 American personnel participated. They included Navy Seals and Army Rangers and other Special Forces. They also included large numbers of the 10,000 American troops stationed in Panama, such as the 193rd Infantry Brigade. Soldiers from the 82nd Airborne Division, 7th Light Infantry Division, and 5th Mechanized Infantry Division also participated. The operation involved simultaneous nighttime airborne operations against twenty-seven objectives throughout the country. Special forces infiltrated key sites shortly before the airborne assaults to take down Panamanian communications and oppose any attempts by Panama to reinforce its forces under attack. The massive simultaneous assault against Panama's 4,400-strong defense forces and its paramilitary forces of several thousand more personnel overwhelmed the latter, surprising them with its ferocity and coordination in the opening hours of battle. Twenty-three Americans died, as did about 125 Panamanian military personnel.[56] Some 200 to 600 Panamanian civilians died as well.[57]

As explained previously, Dupuy's approach begins with a calculation of the power of the two sides. With U.S. forces 22,500 strong, if they are accorded a quality advantage of 3:1 over Panama's military, and assumed to enjoy a 20 percent benefit from surprise, multiplying these factors together gives them a power score of about 80,000. (In his own book, Dupuy uses a somewhat smaller force estimate and a somewhat larger quality advantage, and estimates U.S. power at 75,000.) For Panama, counting about 4,000 paramilitary forces as well as soldiers, it had about 8,500 troops available; the fact that they fought on the defensive and in complex terrain is assumed to give them a doubling of capability score, for a total power of almost 20,000.

So the U.S. power advantage would be about 4, as calculated here (which shows up as the inverse of 4, or 0.25, in the first equation that follows).

American casualties can then be estimated at $(22,500 \times 0.01 \times 0.25) = 56$ a day. Over two days, U.S. casualties would be about 112, with about twenty killed. For Panama, flipping over the power ratio term, we get $(8,500 \times 0.01 \times 4) = 340$. Over two days, Panama's casualties are estimated at 680, with about 170 killed. These results track reasonably well with the actual outcome.[58]

Operation Iraqi Freedom

Before the Iraq invasion of 2003, I used Dupuy-like methods to get a rough sense of how many casualties might be suffered in the course of the conflict, predicting that up to several thousand Americans could die in the struggle. My estimates have, as of this writing in 2008, sadly turned out to be more accurate than seemed initially likely. My roughly correct answer was obtained, however, for partially wrong reasons. That is, my calculations assumed a hard but brief urban fight rather than a protracted guerrilla campaign. These mistakes largely balanced each other out. To put it differently, my instincts about how the war might go were largely wrong, but by being forced to follow the rigor of a model, I wound up with reasonably accurate results nonetheless. That is a testament to the value of models; by forcing one to posit better and worst cases, they can lead to a healthier acknowledgment of uncertainty than many would allow relying solely on personal judgment and intuition.

Here's how the calculation might go. The United States employed almost 200,000 troops in the initial invasion of Iraq—closer to a quarter million, perhaps, counting those in support roles in the broader combat theater as well. Assuming a quality advantage of 3:1 over Iraq's military, and some limited benefit from surprise, the assembled U.S. forces would then have a Dupuy "power value" of about 800,000.[59] If only 100,000 Iraqi personnel are assumed to have fought the United States, with the rest essentially disbanding themselves as the war began, Iraqis would then have a power value of about 200,000 (assuming situational factors from fighting on the defensive that roughly doubled their combat power).[60] That makes the relative power term in Dupuy's casualty equation about 0.25 for the United States and about 4 for Iraq.[61]

If we assume intensive combat, percentage losses per day might be as great as in a smaller conflict like Panama. (Normally, as Dupuy's databases show, larger armies lose a smaller percentage of their forces per day, because large parts of them are well behind front lines at any given time. The relationship varies roughly as the square root of the size of the force—so if an army grows tenfold in size, it will expect to have roughly a threefold reduction in its proportionate loss rate.)[62] So, according to this logic, the daily loss coefficient would remain 0.01.

U.S. losses would then be roughly $(250,000 \times 0.01 \times .25) = 625$ casualties per day, and Iraq casualties would be roughly $(100,000 \times 0.01 \times 4) = 4,000$ per day. For a short war lasting just three days, U.S. casualties could then be about 1,900, of which about 400 might be fatalities. That is how I calculated the lower bound estimate before the U.S.-led invasion of Iraq in 2003.

To estimate the upper bound, I assumed more Iraqis might fight—a quarter million, to be precise. I further assumed the invasion phase of the war could last ten days rather than just three. That led to a U.S. power advantage of just 1.6:1, and an estimate of 1,560 American casualties a day or 15,600 for the whole conflict. That in turn implied about 3,000 dead, the upper range of my estimate.[63]

As of this writing in late 2008, casualties in Iraq have exceeded my upper bound. More than 4,000 Americans have now died in Iraq. In effect, I predicted far too high of a U.S. loss rate—but over far too short a period.

To put it differently, even though I did not do so, the Dupuy equations can be used to model a protracted insurgency/counterinsurgency campaign by adjusting the daily casualty rate appropriately and then playing out the calculation over a long period of time. But in the end, Dupuy's model is not all that useful for understanding or predicting counterinsurgency. For a war in which the size of the enemy is extremely hard to discern, and the enemy's ability to regenerate new forces through recruiting is important, such equations are difficult to use productively since they do not offer insight into such matters. The actual nature of the counterinsurgency combat that ensued (and that continues as of this writing) requires a different type of analytical approach, developed in a subsequent section in the following pages.

Amphibious Assault

The categorizations used earlier—heavy armored combat, infantry combat—are cleaner and neater than warfare often is itself. Many conflicts have elements of both these types of combat. And there are other categories of major military engagement as well. Amphibious warfare is an important such example. Its heyday was undoubtedly World War II, in the Pacific and Normandy and in Italy and elsewhere, though there were also important examples of amphibious operations at Gallipoli, Turkey in World War I, and in Inchon, Korea, and more recently in the Falklands War. The following framework is designed to help gauge the basic feasibility of an amphibious assault operation for a given military balance (not to estimate casualties, which might be best forecast by looking at the individual elements of the battle separately, with tools developed earlier in this chapter). As such, it implicitly provides a way of trying to size a force for an amphibious assault in terms of force planning.[64]

The modern era of precision strike has made some doubt whether amphibious assault really has a future, and the U.S. military's decision not to try to fight ashore in Kuwait in 1991 (largely out of concern over simple Iraqi mines at sea) reinforced the theory that perhaps amphibious operations had become too difficult in current times. But the U.S. Marine Corps, among other organizations and militaries, still takes the amphibious mission quite seriously (maintaining a sealift capacity to put nearly a division ashore in the face of hostile fire, and purchasing amphibious vehicles and tilt-rotor aircraft designed largely, if not primarily, for such operations). There may also be situations where amphibious assault is contemplated for the simple reason that it may be the only way to take a given objective.

How to assess the prospects for successful amphibious assault, as well as the likely course of battle and key determinants of outcomes? Answering these questions requires additional tools beyond those discussed previously for land warfare (though amphibious assault, if at least initially successful, ultimately morphs into land warfare, meaning that equations like those developed earlier can be used to analyze part of the problem). In fact, an amphibious attack is more complex, involving several phases of battle, and the dynamics of aerial attack, naval and air troop movements, and defensive efforts to prevent those movements complicate the analytical exercise. As such, what is needed (prior to using any mathematical formula) is a systematic way of thinking through the respective phases of an operation and asking at each stage if the attacker would possess what would likely be needed to have a chance of success at *each* stage.

To succeed, an invader has generally first needed air superiority, and preferably outright air dominance or supremacy. Second, the attacker has used a combination of maneuver, surprise, and brute strength to land troops in a place where they locally outnumber defenders in troops and firepower. As the previous discussion about data gathered by Trevor Dupuy, the Army's Concepts Analysis Agency, Joshua Epstein, and others shows, it is in principle possible to win battles when outnumbered. But in the case of amphibious assault, that is difficult, since the very act of getting ashore itself provides some warning to the defender—possibly depriving the attacker of some of the surprise that might be achieved more easily on land. Third, the attacker has then generally tried to strengthen its initial lodgment faster than the defender can bring additional troops and equipment to bear at the same location.

If an attacker can do most or all of these things, it has a good chance of establishing and then breaking out of an initial lodgment. As Table 2.1 shows, attackers can succeed without enjoying all three advantages. But in the cases considered here, they did not succeed without at least two of them. For a key modern scenario, an attempted PRC conquest of Taiwan, China might (or might not) gain a large edge in the air; it could not, however, satisfy the other two criteria as discussed more below.

These historical cases, and the framework used here to explain their likelihood of success or failure, probably understates the difficulty of amphibious assault in modern times. For example, the capabilities of modern sea mines outstrip those of mineclearing technologies for the most part, especially in shallow waters.

Even more to the point, the era of precision strike makes it very hard for large assets, notably ships, to approach a defender's shores if the defender has any reasonable combination of prepared defensive positions and advance knowledge of where the attack is to occur. The U.S. Marine

TABLE 2.1
Key Elements in Amphibious Assaults

Case/Attacker	Air Superiority	Initial Troop Advantage at Point of Attack	Buildup Advantage at Point of Attack
Historical Successes			
Okinawa, 1944/U.S.	yes	yes	yes
Normandy, 1944/U.S., allies	yes	yes	yes
Inchon, Korea, 1950/U.S.	yes	yes	yes
Falklands, 1982/UK	no	yes	yes*
Failed Attempts			
Anzio, 1943/U.S. and UK*	yes	yes	no
Gallipoli, 1915/UK, allies	no	yes	no
Bay of Pigs, 1961/Cubans	no	marginal	no

* British forces were outnumbered on East Falkland Island, but they managed to build up their lodgment successfully and move out from it without opposition. At Anzio, although the forces there ultimately contributed to allied victory in Italy in the spring of 1944, their initial objective of making a quick and decisive difference in the war during the previous winter was clearly not met; thus, the operation is classified here as a failure.

Corps, recognizing these developments, has in the modern era placed a premium on maneuver and speed rather than traditional frontal attack—and as noted, the United States chose not to employ an amphibious assault against Iraq in the liberation of Kuwait in 1991.[65]

So how can we ascertain whether, in a possible future war, a given attacker could establish at least two, and better yet three, of the preceding criteria likely to correlate with a successful invasion? The questions on local force advantage are largely a function of the relative logistical capabilities of the attacker and defender to move forces in the face of hostile fire. The first question is slightly more complex.[66] It is not just a matter of who flies the better airplanes or the larger air force, but what effects one can achieve from the skies—and prevent the adversary from itself carrying out. Thus there is some ambiguity in determining if this criterion for possible success is met.

It is easiest to think through these complex issues with reference to a specific example, and given the importance of this scenario, a good case for scrutiny is that of a possible Chinese attack on Taiwan. As is well known, China rejects the idea of Taiwan's permanent separation from the mainland, whereas Taiwan has been seeking to expand its role as a sovereign entity in global affairs, and many Taiwanese feel that the island should be independent. The potential exists for a political crisis that could lead to war—and a war that might ultimately even engage the United States. Even if unlikely, such a conflict is important to analyze given the huge stakes.

Chinese Attempts to Establish Air Superiority

Begin with the matter of trying to establish air superiority (this is the less demanding of the two standards of aerial advantage listed earlier, with air supremacy suggesting a much greater degree of dominance). From China's point of view, ideally it would ground Taiwan's air force early in any war. Otherwise, even the occasional Taiwanese attack aircraft could cause serious losses to PRC ships approaching Taiwan's shores in an amphibious assault, given the accuracy and lethality of modern aerial ordnance, as discussed further in the following. At a minimum, China would need a major edge in the air to make such Taiwanese attacks rare and dangerous for the pilots carrying them out.

China does not have enough of an inherent advantage in air combat capability to achieve such air superiority through dogfighting alone. It would surely need to begin any war with a highly effective surprise strike against Taiwan's air force, which could otherwise take shelter in its many hardened sites on its various airfields, and sneak out for the occasional strike mission against large PRC ships. (It is important to note that if it employed this surprise tactic, China could not start loading and sailing most of its ships towards Taiwan until after the missile and air strikes began, for fear of tipping off Taiwanese and U.S. intelligence about its intentions. In fact, the PRC would do extremely well simply to prepare its air and missile forces for the attack without having those preparations noticed.)

China's ability to carry out such a surprise attack would depend on its missiles and aircraft based near Taiwan, primarily. China has a large ballistic-missile force and a large air force that could be used, among other missions, to attack airfields and planes on those airfields. The missiles are numerous, now totaling about 1,000 in southeastern China near Taiwan. While China's ballistic missiles (as well as its cruise missiles) have been rather inaccurate to date, that is changing.[67] As for planes, China has about 100 airports of all kinds within 600 miles of Taiwan. Approximately 750 military aircraft are normally located at the twenty such airports used by the military.[68]

Taiwan has hardened shelters for most of its fighters. Of course, it would have to keep aircraft in those shelters routinely to survive any surprise attack, or have advance warning of the attack.

Taiwan has another challenge here, too: keeping runways operational. Two to three dozen planes might be needed to shut down a given runway, or a somewhat lesser number in combination with China's more accurate missiles.[69] Taiwan could begin to repair runways after any Chinese strike, assuming it has sufficient runway-repair equipment (and sufficiently hardened maintenance facilities and fuel distribution infrastructure).[70] China could undertake subsequent attack sorties, of course. However, Taiwan's anti-aircraft artillery and SAMs would then be on a high state of vigilance, and the Chinese air force might well lose 5 to 10 percent of their planes on each subsequent sortie, even if able to use standoff precision-guided munitions that allowed them to stay out of the immediate environs of the airfields.

So Taiwan would have a good chance to keep large numbers of aircraft functional after a PRC surprise strike, assuming it was vigilant and careful in day-to-day security operations as well as in its hardening of key facilities.[71] Given the number of shelters it owns, unless it was completely careless, Taiwan should retain about half its air force (perhaps 300 planes) after even a masterful PRC surprise strike.

The Initial Amphibious Assault

The next stage in the battle, amphibious assault, is somewhat simpler to model. It is largely a question of how fast transport ships can be loaded, sailed, and unloaded, as well as of how fast a defender's force can be mobilized and sent to fighting positions.

While the math is simple enough to follow in general, focusing on the China–Taiwan case provides some additional clarity as an example. China has the capacity to transport 10,000 to 15,000 troops with some heavy armor by amphibious lift.[72]

Assume for the moment that China could deploy all of these ships to a single point on Taiwan's shores at once. The question is, could the number of arriving troops rival the number of Taiwanese defenders that would likely be able to arrive at a comparable time?

The defender's response is a function of how soon it would notice the impending assault, how long it would then take to mobilize and activate troops, and how long it would need to position those forces where they were needed. Taiwan would have numerous advantages, starting with the fact that its own reconnaissance capabilities and those of the United States would surely pick up Chinese preparations for the assault well before ships left PRC ports. In fact, by the assumptions of this scenario, it would have been struck with a surprise attack against its airfields, providing obvious warning. The only truly plausible reason that it might not respond promptly has to do with the vulnerability of its command and control network; if key technical systems were knocked out, and if some top political and military leaders were assassinated or kidnapped, Taiwan's response could be delayed. This underscores the importance of redundant command and control as well as clear procedures for maintaining the chain of command when it is under attack. Such matters are hard to model with simple arithmetic; they require more complex and sometimes tedious technical assessments of specific technical vulnerabilities.

Assuming that Taiwan's command and control and communications systems largely survived the opening attack, it would then be rather well positioned to respond. With a large military and a relatively small land mass to defend, as well as a good road network, it would have advantages that not every such defender would enjoy.

To be more concrete, note that Taiwan has:

- 200,000 active-duty ground troops
- 1.5 million more ground-force reservists
- A coastal perimeter of about 1,500 kilometers

Portioning these out, this means Taiwan could deploy roughly 1,000 defenders per kilometer of coastline along all of its shores if it wished. So over any given stretch of ten to fifteen kilometers, a tactically relevant distance (since forces over that distance could all reach incoming Chinese assets with many of their weapons), fully mobilized Taiwanese defense force would be able to deploy as many troops as China could deploy there with all of its amphibious fleet. (An attacker would need to seize a shoreline of roughly that length, to create areas safe from enemy artillery.)[73]

This clearly assumes, however, that Taiwan would make rapid use of its reservists.

The preceding presupposes rapid Taiwanese mobilization, but no advance knowledge by Taiwan about where the PRC intended to come ashore. In reality, unless completely blinded and paralyzed by China's preemptive attacks against airfields, ships, shore-based radars, other monitoring assets, and command centers, Taiwan would see where ships sailed and be able to react with at least some notice. (It is also very likely that, even if it did not immediately send combat forces, the United States would be willing to provide Taiwan with satellite or aircraft intelligence on the concentration of China's attack effort. The United States and Taiwan now have a military hotline, allowing for the possibility that the U.S. global surveillance system could plug holes in Taiwan's own capabilities or replace them after a PRC attack.)[74] Although the Strait is typically only 100 miles wide, Taiwan itself is about 300 miles long, so ships traveling 20 knots would need more than half a day to sail its full length, and could not credibly threaten all parts of the island at once. In addition, amphibious assault troops cannot come ashore just anywhere. Only about 20 percent of the world's coastlines are considered suitable for amphibious assault. On Taiwan's shores, the percentage is even less, given the prevalence of mud flats on the west coast and cliffs on the east.

As a practical matter, then, Taiwan would not need to mobilize all of its reservists to achieve force parity in places most likely to suffer the initial PRC attack. If it could mobilize even 20 percent of its reservists in the days that China would require to assemble and load its amphibious armada and then cross the Strait, it could achieve force parity with China along key beachlines. Thus, China would be unlikely to establish even a local temporary advantage along the section of beach where it elected to try coming ashore.

Taiwan also has at least two airborne brigades that it could use to react rapidly where China attacked.[75] They would allow it to counter China's airlift capabilities (estimated at two to three brigades of paratroopers). China may already be adding a capacity to carry several thousand more light forces with the purchase of thirty Il-76 Russian transport aircraft.[76] However, PRC paratroopers (or troop-carrying helicopters) over Taiwan would be at great risk from Taiwanese fighters, surface-to-air missiles, and anti-aircraft artillery. Paratroopers in fixed-wing transports are particularly vulnerable in situations in which the attacking force does not completely dominate the skies and in which the defender has good ground-based air defenses.[77]

Thus, China does not possess the ability to generate the second element of most successful amphibious attacks as shown in Table 2.1. Its maximum likely deployment of initial forces would at best be comparable to Taiwan's activation of defensive forces in the same landing zone.

Moreover, the preceding analysis does not even include expected attrition to Chinese incoming forces. In reality, such losses would be enormous, both in the initial assault and in subsequent reinforcement operations. Many of the troops crossing the Strait in China's amphibious ships would never make it to land.

Retired Captain Wayne Hughes, Jr. provides a very simple and useful algorithm for understanding why exposed ships at sea are highly vulnerable. As he points out, given the lethality of modern antiship missiles (discussed further in the following), ship defenses are best viewed as filters—taking out a certain percentage of incoming threats—rather than reliable protectors. This is true even before the point where defenses might be saturated with more incoming threats than they are even theoretically capable of simultaneously tracking and engaging. In addition, attackers can often concentrate their fire in salvos of several shots at a time, further complicating the defense's job. Moreover, given modern homing missiles, somewhat less skill is required in firing them effectively than was the case for ordnance in many previous eras of naval combat, making for a high expected accuracy per shot. Also, modern ships are often incapacitated after just a couple shots.

So if, say, eight shots are fired at a given ship, and six are correctly aimed towards the target, only one or two might be intercepted. That means four or five might strike their targets, and often two to four would be enough to incapacitate a given ship—so in this kind of example, the attacked ship would almost surely be sunk.

More generally, Hughes' simple "salvo equation" can be written in simplified form as follows. Assume there are two fleets, with A the number of ships in the first and B the number in the second. B(o) is the initial number of ships for B, and B(t) is the number at a subsequent time, t, after it has taken losses.[78] Those losses are:

$$B(o) - B(t) = [(aA - bB)/s]$$

Here, a is the number of accurate shots fired by A per ship, b is the number of missiles intercepted by B's defenses per ship, and s is the survivability of the typical ship in B's fleet. Put differently, this equation simply says that B's losses are equal to the total number of good shots fired by A, minus the number intercepted by B, then divided by the number of shots typically required to put any given ship out of commission. So if A has five ships each shooting four accurate missiles, for a total of twenty incoming missiles, and B has five ships each capable of intercepting two missiles on average (for a total of ten intercepts), and two hits are needed to sink the typical ship in B's fleet, then five ships will be sunk if the missile shots are well distributed:

$$5 = [(20 - 10)/2]$$

The arithmetic here is clearly very simple, but does give some sense of how ships in modern battle actually perform, and what criteria determine their survival or their demise in battle. Among other things, it shows the importance of early detection (to get in the first salvo), of networking offensively when possible (to ensure an effective distribution of missile firings—not too many and not too few at any enemy ship), and of networking defensively when possible (in the event that networked ships can defend more effectively when working together; if this is not the case, the ships are often better off operating in a more dispersed manner).

Of particular relevance for this case, the formulas can easily be modified to show the effects of attacks on an approaching fleet by a land-based force. The product of the two terms aA can simply be interpreted as the number of successful shots at fleet B, wherever and whatever their origin might be.

Historically in the modern era, more than 90 percent of missiles fired at undefended ships reached their targets (with fifty-four ships sunk or otherwise put out of action with just sixty-three missiles fired). About 68 percent of missiles fired at ships that had defenses but failed to use them properly reached their targets (with nineteen ships sunk or put out of action using thirty-eight missiles). Against ships employing their defenses, about 26 percent of missiles fired reached their mark, with twenty-nine ships incapacitated in one way or another by a total of 121 missiles fired.[79] The data for these cases come from battles before the turn of the twenty-first century. Of course, there is variation from case to case but the overall trends are telling about the difficulty of defense and survivability at sea regardless.

As another way of getting a very rough quantitative grip on the problem, for the Taiwan case or another possible example, consider that the British lost six ships to missiles and aircraft and had up to another dozen damaged, out of a 100-ship task force, in the Falklands War—and that they did not generally have to approach any closer than 400 miles from the Argentine mainland during the conflict.[80] That amounts to an effective attrition rate of roughly 10 to 15 percent during blue-water operations—against an outclassed Argentine military that only owned about 250 aircraft.[81] PRC losses would surely be greater against a foe whose airfields they would have to approach directly, whose air forces would likely retain at least 300 planes even after a highly effective Chinese preemptive attack against airfields, and whose antiship missile capabilities substantially exceed Argentina's in 1982. Taiwan possesses significant numbers of antiship missiles such as Harpoon and its own Hsiung Feng. Weaknesses are evident in Taiwan's capabilities to resist invasion: its air force has focused primarily on air-to-air attack, not antiship operations, and the United States has resisted providing Taiwan with certain attack capabilities out of fear they might be used provocatively. But despite these limitations, Taiwan's panoply of capabilities is considerable, and would be potent even at night or in bad weather. The central point is that, with presumed help from American reconnaissance, and with the advantage of seeing the PRC ships coming and being able to fire at them from relatively safe positions on the shore or from land-based aircraft, Taiwan would have the advantages that the salvo equations show to be so important.

All told, the PRC would likely lose at least 10 percent of its forces just in approaching Taiwan's coasts and fighting ashore each time it attempted another trip. This means that after five trips it could be down to 60 percent (or less) of its initial fleet and after ten trips it could be down to 35 percent.

The situation would get no better over time for China even if it could somehow establish an initial toehold on the island. As the battle wore on, Taiwan's internal lines of communication would help even more, and its ability to reinforce its defensive position at the chosen point of PRC attack would improve further in relative terms, even as China tried to reinforce its attacking legions. By my estimates for this case, China could average deploying no more than 10,000 more troops per day to Taiwan by sea and air combined, in days 3 through 10 of the operation. This assumes that the average ship can do a round trip to Taiwan every other day, on average, including time for loading and unloading. And for days 11 through 20, the average daily flow would be only half that due to ship attrition.[82]

The preceding assumes that China would not be able to seize and protect a Taiwanese port, but that it would instead have to rely on amphibious shipping to an undeveloped area. Taiwan would presumably mount strong defenses near major ports and airfields (and also have the capacity, if truly necessary, to destroy most of the supporting infrastructure to deprive China of the ability to employ it).

By contrast, Taiwan could on average mobilize and deploy another 50,000 troops a day to the location where China was seeking to establish a firm toehold. And its means to continually reinforce would not be nearly as vulnerable to interdiction as would China's.

What if the PRC used chemical weapons in this part of its attack? If it could fire chemical munitions from its ship-based guns, it might be able to deliver enough ordnance to cover a battlefield several kilometers on a dimension within several minutes. China would presumably want to use a nonpersistent agent, like sarin, so its troops could occupy the area within a short time without having to wear protective gear. The effects of the weapons on Taiwan's defenders would depend heavily on whether they had gas masks handy, the accuracy of Chinese naval gunfire, weather conditions, and the speed with which Taiwan could threaten the PRC ships doing the damage.[83] Historical experiences with chemical weapons suggest that China should not expect these weapons to radically change the course of battle in any event. Even in World War I, when protective gear was rudimentary, chemical weapons caused less than 10 percent of all deaths; in the Iran–Iraq War, the figure has been estimated at less than 5 percent.[84] China would need to worry that, if its timing and delivery were not good, its own mobile and exposed troops could suffer larger numbers of casualties than the dug-in defenders.[85] Using chemical weapons could also invite Taiwanese retaliation in kind against China's relatively concentrated and exposed forces on and near the island.[86] All told, this approach would improve China's odds of getting an initial foothold on Taiwan slightly. However, it would not change the fact that Taiwan could build up reinforcements far faster than the PRC subsequently.

Some have raised the possibility that the PRC could use its fishing fleet to put tens if not hundreds of thousands of troops quickly ashore on Taiwan. There are several important reasons not to take this threat particularly seriously, however. First, the ships could not carry many landing craft or much armored equipment. Second, Taiwanese shore-based coastal defense guns and artillery, as well as Taiwanese aircraft, small coastal patrol craft, and mines (not just advanced antiship missiles), might well make mincemeat of many of the unarmored ships, which would have to approach very close to shore in order for the disembarking soldiers not to subsequently drown.[87] Third, given the distances involved, it would be impossible to coordinate the assault very well; the ships would inevitably arrive on Taiwan's shores in ragged staggered formations that would deny PRC troops the benefits of massed attack.[88]

Extracting broader methodological lessons from this case, it is important to break down an attempted amphibious assault into several stages. And in addition to comparing broad force capabilities, such as airpower and tonnage of shipping, it is necessary to evaluate factors like the respective intelligence and monitoring capabilities of each side, the quality of air defenses, the likely care with which command and control facilities and airfields and the like have been prepared against possible surprise attack, and internal lines of communication/transportation for the defender. Being completely confident about the analytical results is difficult, since an ill-prepared defender may be shocked into paralysis or confusion by a sufficiently well-conceived surprise strike. But if analysis shows that one side would effectively need everything to go right even to have a chance at success, and if the other side is capable of maintaining a certain vigilance in its preparations and its day-to-day watchfulness, the defender can probably succeed, given the inherent difficulty of amphibious assault in the missile age.

Some of the preceding challenges facing any amphibious attacker China could address. It could build a great deal more amphibious shipping. It could, and will, continue to improve the accuracy of its missile force and its air-delivered munitions to improve capacities against Taiwan's air force in particular. It could continue to learn more about how to disrupt Taiwan's command and control networks. So Taiwan's defensible position is not guaranteed to endure. It will need to continually harden key assets and devise backup command and control measures, improve runway repair capabilities as well as the strength and number of aircraft shelters, strengthen antishipping missile capabilities, and the like. Nevertheless, its current military position against amphibious assault appears quite robust.

Coercive Uses of Force: Blockades and Barrages

Obviously, not every use of military force amounts to all-out war. Sometimes countries conduct limited attacks for limited purposes. Two cases of modern relevance are considered here: a possible North Korean barrage of Seoul, primarily with artillery, and a Chinese blockade of Taiwan that might or might not be coupled with missile strikes. Of course, analyzing the likely course of any such scenario is a complex matter of political-military analysis; the goal here is simply to explain some of the technical issues involved in assessing how potent any such attacks might be. These two cases dramatize the types of scenarios because Seoul is unusually vulnerable to artillery barrage (being so close to heavily armed North Korea) and Taiwan is particularly vulnerable to missile strikes or even a partially effective, "leaky" naval blockade (being so dependent on shipborne commerce and so close to China).

A North Korean Artillery Barrage of South Korea

What could North Korea do to Seoul and environs by way of bombardment even if it could not plausibly seize the South through outright invasion? If its capabilities were sufficient, South Korea might in theory be intimidated into surrender—or at least into appeasement prior to hostilities, knowing what would happen if war began (and fearful that North Korea might be willing to run the risks, given its extremist regime).

North Korea could, to be sure, seriously harm the South Korean people and economy. About 500 of North Korea's roughly 12,000 artillery tubes are within range of Seoul in their current positions. Most artillery can fire several rounds a minute. Also, the initial speeds of fired shells are generally around half a kilometer per second. That means that even if an ROK (Republic of Korea) counterartillery radar some ten kilometers away picked up a North Korean round and established a track on it within seconds, a counterstrike would not be able to silence the offending DPRK (Democratic People's Republic of Korea) tube for at least a minute (and probably closer to two minutes). On average, such a tube could therefore probably fire two to five rounds, and quite possibly a dozen or more, before being neutralized or forced to retreat fully into its shelter. Some tubes may even be able to fire from protected positions, permitting them to keep up the barrage until they suffer either a near-direct hit by an artillery round or an attack from a laser-guided or satellite-guided bomb.

That means that at least several thousand rounds could detonate in Seoul no matter how hard the allies tried to prevent or stop the attack. The lethal radius of a typical artillery shell is usually thirty to fifty meters (for standard 81mm and 155mm rounds in the U.S. arsenal, with anywhere from seven to twenty-five pounds of explosive). That reference point, as well as historical precedents from conflicts such as the Bosnia war of 1992–1995, suggest that an average round could cause up to tens of casualties and considerable physical destruction. The end result could be up to tens of thousands of civilians wounded, with perhaps one fifth the total number dead. Attacks against Seoul would probably be much worse if they involved chemical weapons.[89]

A Chinese Missile Barrage and Blockade of Taiwan

Two of the most notable ways that China might threaten Taiwan, short of attempting a very risky amphibious assault, are missile attack and naval blockade. The two techniques might also be combined in a single operation.

China has about 1,000 short-range missiles, believed to be equipped with conventional warheads, in the southeastern part of its country near the Taiwan Strait. The missiles could be used in a number of ways, going well beyond what happened in the mid-1990s (when they were fired into the sea near Taiwan). They could be aimed at remote farmland or mountains, to minimize the risk of casualties (if only a few missiles were used, in a strictly symbolic way against such sites, it is plausible no one would be killed, though China could not be sure in advance). They could be aimed very close to land so they would be visible to residents when their warheads exploded. They could also be aimed at the waters just outside ports, or even within harbors, implicitly threatening Taiwan's economy but again without being likely to cause many casualties. If the crisis intensified, successive missile strikes might be aimed closer to shore and closer to cities, with a greater risk of casualties—potentially causing dozens of fatalities in a single strike. Missiles could also be directed at military installations, if China wished to avoid civilian casualties.

The missile option has limits, however. Most of China's ballistic missiles are not yet very accurate, with expected miss distances of 100 meters or more.[90] The types of strikes mentioned here, while perhaps unlikely to cause many casualties, would also be unlikely to achieve much direct and lasting military or even economic effect. And escalation would be problematic for China. Using missiles against cities would be seen as a brutal terroristic act that could do more to unify the people of Taiwan—and the world—against China than to achieve Beijing's war aims. At least, that has been the historical norm when cities have been bombarded in the modern era, be it by airplanes in World War II or ballistic missiles in the Iraq–Iran and Persian Gulf wars.[91] For these reasons, missile strikes might be a logical way for China to begin any use of force, but it would probably need a backup option in case they failed.[92]

Cruise missiles can be much more accurate, and China is obtaining these, too. It may have several hundred with sufficient range to find targets in Taiwan. The warheads on these missiles may be smaller, and their likelihood of being shot down much greater, but this threat may be, on balance, somewhat more militarily meaningful. Still, perspective is needed; the United States has frequently used cruise missiles in modern war, and often used several hundred in a given conflict, but has never come close to achieving widescale military objectives with such missiles alone. In modern wars, it has typically had to deliver many thousands of precision bombs to achieve its goals.[93]

That is where a naval blockade could offer appeal to China. It could be "leaky" and still directly threaten Taiwan's economy. To do so, it need not physically stop all ship voyages into and out of Taiwan. It would simply need to deter enough ships from risking the journey that Taiwan's economy would suffer badly. The goal would be to squeeze the island economically to a point of capitulation. This solution could seem quite elegant from Beijing's point of view—it could involve only a modest loss of life, little or no damage to Taiwan itself, more terror than harm suffered by the people of Taiwan, and the ability to back off the attack if the United States seemed ready to intervene or if the world community slapped major trade sanctions on China in response. Additionally or alternatively, the capabilities needed to carry it out, most notably submarines as argued in the following, could also help deter and complicate any American naval intervention on behalf of Taiwan.[94] How to analyze such a blockade systematically and quantitatively, when its principal goals might be qualitative and psychological?

A Chinese blockade could take a number of forms. Militarily speaking, the least risky and most natural approach would simply attempt to introduce a significant risk factor into all maritime voyages in and out of Taiwan by occasionally sinking a cargo ship with submarines or with mines China laid in Taiwan's harbors.[95]

Using airplanes and surface ships would put more of its own forces at risk, especially since it could not realistically hope to eliminate Taipei's air force with a preemptive attack (though airpower might be used in a hit-and-run raid, especially as an initial strike before Taiwan's defenses were fully alerted). A blockade using planes and surface ships would also be rather straightforward for the United States to defeat quickly. It benefits from superior scouting/reconnaissance abilities at sea; in addition, the lethality of modern naval weapons means that the side able to muster an effective attack first is increasingly in the dominant position.[96]

To be sure, a blockade would be challenging and dangerous for China to pull off. Perhaps the greatest worry for Beijing would be its likely inability to distinguish one country's merchant ship from another's. But if Beijing announced to the world that those shipping towards Taiwan were aiding and abetting its enemy, and gave fair warning, it might consider itself to have done enough to warrant attack against any vessel not heeding its demands. Moreover, it might offer countries the option of first docking in a PRC port for inspection (if it decided to allow humanitarian goods through, for example, or ships from certain friendly countries but not others) and then being escorted safely to Taiwan. Since this strategy might require it to sink only a few ships to achieve the desired aims, even in a worst-case scenario it might believe it was threatening the lives of only 100 to 200 commercial seamen. Given the perceived stakes involved, Beijing could well consider this a reasonable risk.

Most of China's submarines do not have antiship cruise missiles or great underwater endurance. However, the PRC submarine force is steadily improving. Even today, Chinese subs have adequate ranges on a single tank of fuel—typically almost 10,000 miles—to stay deployed east of Taiwan for substantial periods. Although their ability to coordinate with each other and reconnaissance aircraft is limited, that might not matter greatly for the purposes of a "leaky" blockade. Even if picking up commercial ships individually by sonar or by sight, such submarines could maintain patrols over a large fraction of the sea approaches to Taiwan. It could take Taiwan weeks to find the better PRC submarines, particularly if China used them in hit-and-run modes.[97] Given the lethality of modern torpedoes and cruise missiles (for any PRC submarines carrying the latter), the existence of these submarines in important waterways near Taiwan would constitute a very major threat.

To break the blockade, the basic idea for the United States and Taiwan would probably be to deploy enough forces to the Western Pacific to credibly threaten the following type of operation. The magnitude of this operation shows how hard it is to reliably defeat even an imperfect blockade.

To break the blockade, the United States and Taiwan would have to set up a safe shipping lane east of Taiwan. In addition, they would have to heavily protect ships during the most dangerous part of their journeys when they were near the island. To carry that mission out, the United States, together with Taiwan, would need to establish air superiority throughout a large part of the region, protect ships against Chinese submarine attack, and cope with the threat of mines near Taiwan's ports.

The anti-submarine warfare (ASW) effort could have multiple aspects. The United States would surely be tempted to deploy its own attack submarines as close as possible to China—certainly in the Taiwan Strait, maybe just outside PRC ports. This approach would provide American submarines a good prospect of destroying PRC subs at their source, before they were in a position to fire on commercial shipping (or U.S. aircraft carriers) in more distant waters. However, this type of ASW would be extremely delicate strategically, especially if it involved attacks in Chinese territorial waters.

Whatever happened near Chinese shores, there would surely be additional layers of American ASW further out to sea. The convoys sailing to and from Taiwan would need protection. American ships, primarily ASW frigates, would accompany convoys of merchant ships as they sailed in from the open ocean waters east of Taiwan. These convoys might form a thousand miles or more east of Taiwan, and enjoy armed protection from that point onward as they traveled to the island and later as they departed. The frigates would use sonar to listen for approaching submarines, and for the sound of any torpedoes being fired. Some ships would be larger destroyers or cruisers, such as those equipped with advanced Aegis radars, to detect any use of cruise missiles and attempt to defend against them.

The United States would probably deploy significant numbers of surface combatants and airplanes like P-3s to the region for this mission. Some would help protect U.S. aircraft carriers, of which at least four would likely deploy east of Taiwan to establish air superiority in the event of any conflict. Others would provide additional protection to merchant ships or mine warfare vessels as they operated near Taiwan's shores. U.S. minehunters and minesweepers would operate near Taiwan's ports and the main approaches to those ports. Land-based or ship-based helicopters might assist them. So might robotic submersibles deployed from ships near shore.

If China then used its submarines in attacks on shipping, or if direct hostilities began in another way, the United States would almost surely begin to actively search for and fire upon Chinese submarines as a matter of normal operations. Any Chinese submarine wishing to fire at a merchant ship or aircraft carrier would then first have to run quite a gamut. It would have to evade submarine detection as it left port, avoid any open-water search missions that the United States and Taiwan established, and then somehow penetrate the defensive ASW perimeter of whatever convoy it was attacking as it approached its target. To survive the overall engagement and return to port, it would then need to successfully negotiate all of this in the other direction.

During the Cold War, the effectiveness of ASW operations was commonly assessed at 5 to 15 percent per "barrier" (Cold War barriers at that time were more linear and literal perimeters than would be likely here, but the fact remains that Chinese subs would have to survive perhaps three types of pursuers at three different parts of their journey to or from home base.) By those odds, the typical Chinese sub would do well to survive for two or three roundtrip missions from base.[98] But it might succeed in getting off several shots against valuable surface ships before meeting its own demise.

The vulnerabilities of all ships (including U.S. Navy ships) to attack are amplified in shallow littoral waters. In the open blue waters of the oceans, the U.S. Navy can generally detect enemy ships or aircraft long before they are close enough to strike. But in shallow waters, shore-launched antiship missiles are a threat, as are weapons fired from aircraft or ships that dart out from a country's coastal regions. A similar conclusion applies to the threat from submarines and torpedoes. In the open oceans, the U.S. military can rely on sonar (from aircraft, fixed underwater arrays known as SOSUS, and ships) to get a good sense of the approach of enemy submarines. Sonar is relatively predictable in deep waters; moreover, any ship approaching a U.S. vessel would have to travel a great distance to reach it in such a location, offering multiple opportunities for detection. By contrast, shallow waters are complex sonar environments, where sound waves bounce back and forth in multiple and unpredictable directions. This makes ambush a real worry, especially for the mine warfare vessels and surface ships that would have to escort commercial vessels all the way into Taiwan's ports.[99]

To be sure, the United States Navy would not deploy most of its assets near China all at once. However, China still might think that a quick strike that sank a carrier and killed hundreds or thousands of Americans would cause Washington to waver in its future commitment to the defense of Taiwan. As naval analyst Captain Wayne Hughes, Jr. argues about the nature of naval combat:[100]

- Defense is usually weaker [than offense, unlike land battle].
- Defensive power is solely to gain tactical time for an effective attack or counterattack.
- When two competitive forces meet in naval combat, the one that attacks effectively first will win.

Hughes also argues that there is perhaps less friction in battle at sea than on land, and that many naval engagements occur in which decisive results happen relatively quickly, in contrast to the norm for war on land.

Technology trends could put ships at even greater risk today. Hypersonic antiship cruise missiles are becoming more common and are extremely difficult to defend against, even for high-performance U.S. Navy ships with advanced Aegis radar systems. The ranges of PRC cruise missiles are now reaching or exceeding 150 miles. To make these weapons more effective, China can be expected to try to improve its targeting and communications systems, too. For example, it is putting into orbit more satellites capable of detecting large objects on the oceans.[101] With the information from satellites, guidance systems on the cruise missiles could then guide them to the vicinity of their targets, where terminal seekers on the missiles themselves could finish the navigation job.[102] The only truly reliable way to protect ships against such threats is to minimize the number of missiles that can be fired at them by depriving the missile-carrying aircraft or ships of proximity to their would-be targets. The United States would have a good chance to do this successfully against China, but the PRC's submarines would complicate the task and cause a real risk of significant American losses in the course of battle.

Nuclear Exchange Calculations

Although more a vestige of the Cold War than a focus of current defense planning, nuclear "exchange calculations" are still worth understanding. The physics underlying them remain relevant even if the nuclear superpower dynamics that led to their centrality in defense circles now seem (mostly) anachronistic. More to the point, perhaps, certain aspects of these calculations could be relevant in considering other issues, such as possible nuclear exchanges between regional powers or between a regional power and the United States—and perhaps most importantly, arms control arrangements designed to lower the probability of such nuclear attacks.

These exchange calculations focused on a nuclear-armed "triad" of forces that the Soviet Union and the United States maintained during much of the Cold War—and that Russia and the United States maintain today, at lower force levels. Smaller nuclear powers typically aspire to a similar distribution of nuclear forces across different types of platforms and delivery vehicles. The triad is made up of intercontinental ballistic missiles (ICBMs), historically placed in the ground in hardened concrete silos or made mobile to complicate the enemy's targeting; submarine-launched ballistic missiles (SLBMs) on submarines that can be deployed at sea to enhance their survivability against attack; and bomber-launched nuclear munitions (bombs or cruise missiles), delivered by planes that can if necessary be placed on runway alert (or even be maintained with some fraction constantly airborne), again to enhance survivability.

An essential goal of nuclear force planners for much of the Cold War was to ensure that their own country's forces could survive any plausible attempt by the enemy to disarm them in an all-out surprise first strike. At the same time, force planners also sought, especially on the American side early in the Cold War, to maximize their capacity for denying the adversary the very survivable second-strike force that they knew necessary for their own country. Worries about ensuring a survivable deterrent drove the two states to create triads in the first place, to harden silos (or, later, build mobile ICBMs that could not be easily pinpointed by the potential enemy), build quieter and quieter (hence harder and harder to find) submarines, put bombers on runway alert at interior bases, and build early detection radars and satellites to watch for any possible "bolt from the blue" attack by the other.[103]

In assessing whether they had achieved their core defensive goal of creating survivable forces (and in also assessing whether they could destroy much of an enemy's nuclear assets through "counterforce" strikes), nuclear planners employed exchange calculations. The basic logic went like this. Assuming one side might be willing to launch an all-out zero-warning attack on the other, how well might it do?

The details of the calculations were based on the following basic concept. First, ICBM silos would be attacked by the other side's most accurate and lethal ballistic missile warheads, since they are difficult to destroy. Typically, two warheads would be used against each silo, to account for the imperfect reliability and accuracy of the incoming warheads. (It was further assumed that by the time one side's bomber forces could penetrate the airspace of the other, the attacked side would have had six to ten hours of warning—allowing it to launch its own ICBMs before they could be attacked. Thus it was generally assumed that bombers would not be the appropriate delivery vehicle for attacking silos.) Submarines carrying SLBMs but located in their ports would be destroyed (a fairly straightforward proposition, since submarines are not very hard relative to ICBM silos); deployed submarines would be hunted down by the other side's attack submarines (probably something that only the United States ever had the capacity to do). Bomber bases could be barraged, again by warheads from ballistic missiles. Any bombers that had managed to become airborne before the barrage occurred would have to be shot down by air defenses. This was the basic picture of any first strike.

Many assumptions, besides the previously noted point about bombers not being well suited to attack ICBM silos, were built into these models—and not all of them were guaranteed to be right. One assumption was that neither side would launch its ICBMs and SLBMs in the fifteen to twenty minutes it might have available, between when it detected the other side's massive launch of missiles and when the incoming warheads would begin to detonate. A second assumption was that the command and control and communications systems of each side would survive initial attacks and be capable of ordering and coordinating retaliation; otherwise, having one's forces survive might be of little benefit.[104] A third assumption behind the way nuclear force planning was done (if not the exchange calculations themselves) was that keeping forces on alert would not lead to a high risk of accident that would trump the strategic benefits.[105] A fourth assumption was that any conventional war that preceded the nuclear exchange would not lead to confusion about whether a nuclear attack had begun (when, in fact, it had not).[106] Again, if incorrect, this assumption would not invalidate the exchange calculation methodology per se, but it would call into question the logic of the prevailing force planning paradigms.

Some of these assumptions can be assigned a probability, or be analytically assessed; others are more intangible. For example, the idea that neither side would launch on warning was hardly an inevitability and may well have been wrong at many times during the nuclear age—but it still was a useful simplifying way to assess the *maximum* vulnerability of a given side's forces. The assumptions about the survivability of command and control, by contrast, could in large measure be subject to detailed technical examination of the effects of nuclear blasts on various technologies integral to the command, control, and communications effort. In fact, many steps had to be taken to ensure survivable command and control since technical studies often revealed vulnerabilities—with both superpowers naturally trying to diagnose and repair their respective vulnerabilities before they were recognized by the other. The hope that accidents would not occur could be evaluated either by appeal to organization theory (trying to find analogies for how well large organizations had maintained strong safety records) or by examination of the history of the nuclear weapons business itself, once enough years had passed to create a reasonable dataset of experiences with airborne bombers, at-sea submarines, and the like.

In addition to such assumptions, calculations also were done with certain simplifications. For example, it was assumed that two warheads could be detonated near a single silo, without the shockwave, x-rays, or debris cloud from the first destroying the second (a process known as "fratricide").[107] It was further assumed that missiles would perform the same way on flight trajectories over the North Pole as they had on test ranges. It was also assumed that, while ballistic missiles might individually fail, there would be no systemic problem with warheads detonating—even though those warheads would never have been tested under truly realistic conditions before.

Keeping these potentially flawed assumptions in mind, this is how the exchange calculations then proceeded for some typical weapons involved. Assume, say, Soviet SS-18 missiles (with ten warheads each) attacking American Minuteman missile silos. The reliability of any SS-18 missile itself was estimated at 85 percent. The average miss distance (or "circular error probable") of a warhead launched by an SS-18 was estimated at 150 meters. Since the warhead yield was estimated at 500 kilotons, corresponding to a "lethal radius" of 290 meters against American ICBM silos (given their specific level of hardening, estimated at a resilience of up to 2,000 pounds per square inch of overpressure from a blast wave), that meant most SS-18 warheads would destroy the silo they were aimed at (since most would land within 290 meters of their aim points, given their typical miss distance of only half that distance). The main hope for the silo, by this methodology, would be that any and all missiles launching warheads at it would fail. Assuming two different missiles were used to launch the two warheads directed at each silo—a practice that complicates targeting, of course, but hedges against the failure of any given rocket—this implies a 95 percent chance of any given U.S. Minuteman missile being destroyed.[108]

For submarines, the math was simple. All SLBMs in port would be destroyed; all U.S. SLBMs deployed at sea would survive. Soviet-deployed SLBMs, for their part, might or might not survive depending on how well U.S. attack submarines and other antisubmarine warfare capabilities were able to do their jobs in finding the Soviet subs carrying the SLBMs.

For bombers, all bombers on the ground would be assumed to be destroyed, all those well into the air before the attack would survive (and then stand some chance, perhaps 50 to 90 percent, of successfully penetrating enemy air defenses as they approached their targets). Only for those trying to get off the ground as warheads started to explode around them would the math be complicated, as the attacker might have sought to barrage the airspace around the runways, creating enough overpressure to knock planes only just getting off the ground out of the sky. The mathematics of this process are not discussed in detail here. Suffice it to say that a warhead of roughly 500 kilotons' yield could likely destroy aircraft out to about three to five kilometers distance (though there is considerable uncertainty in this estimate and it could easily vary by a factor of two or even more, either way).[109] However, since planes travel fast and can vary their flight direction and altitude quickly, a barrage scenario is a complex one to evaluate.

Typically, exchange calculations showed that either superpower might retain 20 to 50 percent of its initial forces after an enemy first strike during most of the latter decades of the Cold War. Each would still have retained a great deal of redundancy or "overkill" in its nuclear forces, clearly. The situation clearly might not be so simple, however, for smaller nuclear weapons, and the potential for high vulnerability could be much greater.

Sizing Stabilization and Peacekeeping Forces: The Case of a Collapsing Pakistan

How should peacekeeping or stabilization missions be sized and structured? This is a difficult question, because there is no simple one-size-fits-all formula. In cases where parties to a conflict are truly exhausted by fighting and committed to peace, or cases where available weaponry is limited, for example, modest numbers of troops may suffice. Then again, it is this logic—or this sort of hopefulness—that may have led to the carelessness with which the United States prepared for the occupation of Iraq. In cases where the risk of failure is truly not acceptable, what would cautious defense planning metrics say about necessary troop levels?

This section of the chapter employs a concrete example to motivate and illuminate an otherwise rather dry set of calculations. The case in point is Pakistan. Of all the military scenarios that would undoubtedly involve the vital interests of the United States, short of a direct threat to its territory, a collapsed Pakistan ranks very high on the list. This is not the usual peacekeeping scenario, to be sure, and it is (one hopes) extremely unlikely to occur. But its potential importance makes it a good subject for discussion and analysis—and Pakistan's sheer size underscores the tyranny of the arithmetic that drives this type of force planning. This is because the most notable characteristic of this type of military analysis is the direct relationship between the size of the indigenous population in question and the resulting necessary size of the stabilization force.

The combination of Islamic extremists and nuclear weapons in Pakistan is extremely worrisome. Were parts of Pakistan's nuclear arsenal ever to fall into the wrong hands, al-Qaeda could conceivably gain access to a nuclear device with terrifying possible results. Another quite worrisome South Asia scenario could involve another Indo–Pakistani crisis leading to war between the two nuclear-armed states over Kashmir.[110]

The Pakistani collapse scenario appears unlikely given that country's relatively pro-Western and secular officer corps.[111] But the intelligence services, which effectively created the Taliban and have also condoned, if not abetted, Islamic extremists in Kashmir, are less dependable. Plus, the country as a whole is sufficiently infiltrated by fundamentalist groups—as the attempted assassinations against former President Musharraf (and other evidence) make clear—that this terrifying scenario of chaos cannot be dismissed.[112]

Were it to occur, it is unclear what the United States and likeminded states would or should do. It is very unlikely that "surgical strikes" could be conducted to destroy the nuclear weapons before extremists could make a grab for them, since it is doubtful the United States would know their location, and it is at least as doubtful that any Pakistani government would countenance such a move, even under duress.

If a surgical strike, series of surgical strikes, or commando-style raids were not possible, the only option might be to try to restore order before the weapons could be taken by extremists and transferred to terrorists. The United States and other outside powers might, for example, respond to a request by the Pakistani government to help restore order. But given the embarrassment associated with requesting such outside help, it might not be made until it was almost too late. Hence, such an operation would be an extremely demanding challenge, but given the stakes, there might be little recourse than to attempt it.

What could this type of mission entail? The international community might team with Pakistani forces to help defeat any insurrection. Or it might help protect Pakistan's borders, making it hard to sneak nuclear weapons out of the country, while providing only technical support to the Pakistani armed forces as they tried to put down the insurrection. All that is sure is that, given the enormous stakes, the United States would literally have to do anything it could to prevent nuclear weapons from getting into the wrong hands.

Should stabilization efforts be required, the scale of the undertaking could be breathtaking. Pakistan is a very large country. Its population is about 170 million, or six times Iraq's. Its land area is roughly twice that of Iraq; its perimeter is about 50 percent longer in total. Since a U.S. force of some 140,000 (as part of a broader international force of about 165,000) was ultimately determined to be inadequate to handle the Iraq mission, stabilizing a country of Pakistan's size might seem to require more than a million foreign troops.

To be more precise, according to optimal counterinsurgency doctrine, stabilizing a given region should ideally involve twenty to twenty-five military or police personnel for every thousand indigenous citizens.[113] Put differently, that is at least one peacekeeper for every fifty citizens, implying in Pakistan's case up to three million security force personnel. Such ratios have sometimes been achieved, notably in post–World War II Germany, in Bosnia and Kosovo in the 1990s, and even in Somalia in the early 1990s for a time.[114] Even a more modest approach that accepts somewhat greater risk suggests deploying one peacekeeper for every one hundred members of the civilian population.[115] Only in relatively benign post-conflict environments with an exhausted or thoroughly defeated population (Japan in 1945, Namibia in 1989, El Salvador in 1991, Mozambique in 1993) has it been possible to make do with only one peacekeeper for every 200 or more citizens of the host nation, still implying almost one million forces for Pakistan's population.[116]

This arithmetic quickly shows that no international force could do the job on its own. The world does have more than 20 million soldiers across all of its militaries, but obviously most are unavailable at any given moment for peace operations and stabilization missions and most countries are unable to deploy and sustain substantial numbers of forces abroad in a timely fashion even when they choose to try. The international community has reached new heights in recent years deploying some 150,000 or more total troops in various peace operations (including those under U.N. auspices, as well as under organizations like NATO or the African Union), and a comparable number in Iraq. Its maximum capacity is no more than twice that aggregate amount, or a bit more than half a million troops, even assuming an all-out American effort. That size force could only be deployed over a period of many months—and longer still for some inland locations, given the logistical limitations of many militaries and the corresponding need to establish transportation and support for them (through contractors or other means).[117]

Presumably even in a scenario of a gradually fraying or dissolving Pakistan, some fraction of the country's own security forces would remain intact, able, and willing to help defend their country. By this logic, the international force would deploy only at the request of the host government—which would, in any case, have to continue to handle most of the challenges of the job itself for months as the international force deployed.

Pakistan's military numbers 550,000 Army troops, 70,000 uniformed personnel in the Air Force and Navy, another 510,000 reservists, and almost

300,000 gendarmes and Interior Ministry troops.[118] Police forces are also substantial. But if some substantial fraction of the military broke off from the main body, say a quarter to a third, and were assisted by extremist militias, it is quite possible that the international community would need to deploy 100,000 to 200,000 troops to ensure a quick restoration of order. Given the need for rapid response, the U.S. share of this total would probably be a majority fraction, or quite possibly 50,000 to 100,000 ground forces. Obviously, this calculation is notional and illustrative, not precise; in a given scenario, the numbers could be much different.

Of course, proper force sizing does not guarantee a successful outcome. History contains numerous examples of failures even when theoretically adequate forces were deployed, as with the French experience in Algeria from 1954–1962 (about forty troops per 1,000 indigenous citizens) and the U.S. experience in Vietnam (about eighty-five troops per 1,000 Vietnamese). Clearly a host of other factors impinge on a mission's prospects, including the training and quality of the counterinsurgency/stabilization forces, the perceived political legitimacy of their mission, and the availability of sanctuaries in neighboring states for insurgents. But being wary of the importance of force numbers is one thing; totally ignoring their relevance is something else. Without adequate forces, it is impossible to protect the population, and counterinsurgents often fall back on excessive use of firepower to compensate for their lack of presence.[119]

Assessing Counterinsurgency and Stabilization Missions

How to tell if a counterinsurgency war is being won? Clearly sizing the force correctly for a stabilization mission, as discussed earlier, only addresses the question of getting the inputs roughly right. What about results on the ground?

In conventional warfare, deducing trends is fairly obvious—if not to predict outcomes, at least to discern who is "ahead" at a given moment. Movement of the front lines, industrial production of war matériel, and the logistical sustainability of forces in the field provide fairly obvious standards by which to assess trends. But counterinsurgency and stabilization operations—like the one in Iraq (used here as an important recent example)—are different, and more complex. How do we measure progress in such a situation?

This is a hard challenge because it is easy to misuse and abuse metrics. In Vietnam, for example, the United States was convinced that there would be a "crossover point" in attrition of the Viet Cong. If U.S. military forces could manage to kill enough of them, say 50,000 a year, their recruiting efforts would not be able to keep pace, and combined American and South Vietnamese forces would ultimately prevail. This was based in part on the conviction that successful counterinsurgency requires ten government soldiers for every insurgent—a simplifying assumption that, while partly validated by history, gave American policymakers too much confidence that a given number of U.S. troops could produce victory. That approach led to General Westmoreland's famous search-and-destroy concepts for ground operations. It resulted in a focus on massive firepower that killed huge numbers of innocents and failed to achieve its military objective as well. The conviction that the Viet Cong needed hundreds or thousands of tons of supplies daily led to additional bombing of the Ho Chi Minh trail and ultimately Cambodia—again to no avail as it turned out that the Viet Cong in South Vietnam needed little outside help.[120]

The U.S. focus on supporting a government with strong anti-communist credentials led to dependency on a corrupt regime with limited legitimacy among its people. American hopes about sparking GDP growth in Vietnam were dashed because the country's economic successes accrued only to a small fraction of the population. Finally, Washington's focus on enlarging and equipping South Vietnamese security forces could not compensate for their qualitative deficiencies.[121]

The experience of successful counterinsurgency and stabilization missions in places such as the Philippines and Malaya (now Malaysia), by contrast, tends to place a premium on tracking trends in the daily life of the typical citizen. How secure are they, and who do they credit for that security? How hopeful do they find their economic situation, regardless of the nation's GDP or even their own personal wealth at a moment in time? Do they think their country's politics are giving them a voice?[122]

The Marine Corps tended to focus on these metrics in Vietnam, and developed an approach called the Combined Action Program (CAP) to help protect the population in "ink spots" that would gradually expand with time. In fact, the Marine CAP concept applied more broadly would have led to fewer overall American forces than were actually deployed, suggesting that the ten-to-one rule was NOT the optimal way to gauge U.S. force requirements. But the Marine Corps did not carry the day with this concept for the U.S. military overall.[123] The U.S. military finally moved towards this type of thinking in Iraq—but, in general, not until 2007.[124]

However, tracking trends in the well-being of a population is extremely difficult. Many considerations enter into this question. This helps explain why the Iraq Index at Brookings has included more than fifty key indicators since we began it in late 2003. Further complicating matters is that information is often unreliable. In Iraq, data has been conspicuously poor about the unemployment rate, the crime rate, and trends in the availability of water, sewage treatment, and health care. Finally, in gauging whether a given strategy is working, some metrics may be leading indicators of success (or failure), whereas others may lag. Is there a way to make sense of the cacophony of data?

Once a counterinsurgency has made major strides forward, as Iraq had by this writing in late 2008, it is easier to tell a clear story from the data. Beginning with civilian fatalities from violence (of all forms), perhaps the ultimate indicator of stability, Iraq's rate of violence had declined by 80 percent relative to the 2006 peak and was even lower than in the years 2004/2005. Moreover, it was continuing to decline even as the U.S. surge of forces ended and America reduced its combat brigade strength from twenty to fifteen. With U.S. troop fatality rates down by 60 to 80 percent by mid-2008 as well, and Iraqi Security Force casualties reduced by more than half, too, the overall trajectory of the war was fairly unambiguously good—just as it had been unambiguously bad in 2006.

More complicated is to assess the dynamics of such a war as the situation is changing but before trends are dramatic. This was essentially the situation in Iraq in 2007. As the surge brigades arrived from the United States over the first half of the year, and Iraqi security forces continued to grow at a rate of almost 10,000 uniformed personnel a month, new operations were initiated and the battlefield changed substantially. Nonetheless, violence remained very high; it was not until the latter half of the year that the situation markedly improved throughout much of the country. The U.S. political debate over the surge was meanwhile quite acute, and Congress was considering cutting off funding for the war even as the surge began during the first six to eight months of the year. What indicators could it have looked to, during this transitional time, to determine whether it was worth keeping American forces involved in the fight?

It is difficult to create a clearly prioritized list because leading and lagging indicators could vary from one conflict to another. In Iraq, reductions in U.S. and Iraqi security force casualties lagged because the surge led to heavy fighting in parts of the country as Shiite militias, al-Qaeda in Iraq extremists, and others battled back for a time. Improvements in basic economic-quality-of-life indicators, such as numbers of children in school, the quality of health care, the unemployment rate, and the availability of potable water and electricity continued to lag even in late 2008.

In other cases, the nature of the information may be inherently difficult to interpret. For example, "body counts" of killed enemy combatants may indicate progress—as long as the right people are being killed. But if innocents, or would-be allies, are killed by government forces, the effect can be negative. The latter dynamic probably existed in Iraq in 2004 through 2006; the former appears to have been established by 2007, but body counts themselves would not show the change.

By contrast, some indicators were more promising. Ethnic cleansing rates declined by mid-2007. The numbers of extremist leaders purged from the Iraqi Security Forces and other Iraqi government positions increased quite a bit (though it took a while to be confident that their replacements had higher integrity). Increases in the number of Iraqi security forces taking primary responsibility for local security were also encouraging. But we did not yet know for sure, in 2007, if they would be able to do so in the ethnically mixed neighborhoods in and around Baghdad, Mosul, and Kirkuk or in particularly tense regions like Basra and Sadr City. Only in the spring of 2008 were improvements in Iraqi forces validated by battlefield progress in such places.

Political progress in Iraq was slow through most of 2007, though it picked up as the year unfolded. Knowing how to gauge political progress is hard. It is not a matter of meeting specific "benchmarks" so much as creating a spirit of nonviolent politics and compromise, so that future disputes will be settled in the halls of parliament rather than on the streets or battlefields. Benchmarks are ways of gauging possible progress towards this attitude, but no more than that, and as such must be taken with grains of salt. In 2007, the main progress was in purging extremists from government jobs (as noted earlier), in Baghdad sharing more resources with Iraq's eighteen provincial governments, and in deploying Iraqi forces to places where they could support the U.S.-led surge.

My own confidence in the new strategy grew greatly after a trip to Iraq in mid-2007, but the data were not totally conclusive at that point. It was the combination of some encouraging data trends with a general sense that the United States and Iraq had developed a proper counterinsurgency and stabilization strategy that gave me (and colleague Kenneth Pollack) confidence—underscoring again that quantitative metrics must often be married with military and strategic judgment to reach bottom-line policy judgments in this field. The science of war only goes so far.

By early 2008, things had improved much more. Progress was evident in a new pensions law, in amnesty legislation for some militia fighters, in an improved de-Baathification statute, and in a provincial powers act. Jason Campbell and I hazarded an estimate that Iraq's politics merited a "score" of roughly 5 on a scale of 0 to 11 (using eleven benchmarks for these purposes). This was an imprecise approach, subject to future revision, but seemed the best way to gauge progress on issues that were both inherently important and topical within Iraq. Since then, progress has varied. We accorded the Iraqis 0 for resolving the logjam over the disputed city of Kirkuk's future, for creating a permanent hydrocarbons law, and for passing a provincial election law in mid-2008, but the situation was still unmistakably improved relative to 2007 or earlier periods. By the end of 2008, our "score" for the Iraqi political system was a 7.

Other cases would require different frameworks. In fact, Iraq itself will require a dynamic approach to assessing political progress in the months and years ahead, given the nature of this business. In the counterinsurgency and nation-building business, no formula can determine whether success or failure is a certainty. But a careful and broad use of multiple metrics can help detect key trends, diagnose problems, and test various theories about whether net progress is occurring.

The following examples help illustrate the uses of some of these modeling and analysis methods. Some focus on current potential scenarios, while others are more generic or historical in character. Some employ a more quantitative approach than others, but all seek to use a systematic and step-by-step approach to answering the questions posed—that is what I mean when referring to combat modeling.

QUESTION 6: What scale of military operation might be needed to secure Kashmir or Congo or Indonesia?

ANSWER: Like the earlier case study on the possible collapse of Pakistan, this is a question of force planning rather than combat simulation or modeling.

Consider first a scenario pitting Pakistan against India over Kashmir. It is highly doubtful the United States would ever wish to actively take sides in such a conflict, allying with one country to defeat the other. Its interests in the matter of who controls Kashmir are not great enough to justify such intervention, and no formal alliance commitments oblige it to step in. Moreover, the military difficulty of the operation would be extreme, in light of the huge armed forces arrayed on the subcontinent and the inland location and complex topography of Kashmir. In addition to the numbers associated with Pakistan, India's armed forces number 1.3 million active-duty troops, and feature such assets as 4,000 tanks, sixteen submarines, and about 600 combat aircraft (defense spending in the two countries was roughly $5 billion in Pakistan and $25 billion in India in 2006, respectively).[125]

However, there are other ways in which foreign forces might become involved. If India and Pakistan went up to the verge of nuclear weapons use, or perhaps even crossed that threshold, they might consider what was previously unthinkable to New Delhi in particular—pleading for help to the international community. For example, they might agree to allow the international community to run Kashmir for a period of years. After local government was built up, and security services reformed, elections might then be held to determine the region's future political affiliation, leading to an eventual end to the trusteeship. While this scenario is admittedly a highly demanding one, and also unlikely in light of India's adamant objections to international involvement in the Kashmir issue, it is hard to dismiss such an approach out of hand if it seemed the only alternative to nuclear war on the subcontinent. Not only could such a war have horrendous human consequences—killing many tens of millions—and shatter the tradition of nuclear non-use that is so essential to global stability today. It could also lead to the collapse of Pakistan, and thus the same types of worries about that country's nuclear weapons falling into the wrong hands discussed earlier in these pages.

What might a stabilization mission in Kashmir entail? The region is about twice the size of Bosnia in population, half the size of Iraq in population and land area. As noted in the earlier section, according to optimal counterinsurgency doctrine, stabilizing a region of 10 million would probably require 200,000 to 250,000 troops.[126]

However, the idea of using twenty to twenty-five peacekeepers per 1,000 civilians is a historically based rule of thumb, not a binding requirement. To be sure, it is dangerous to ignore rules of thumb in favor of hunches or personal preferences, as has sometimes been done in force planning for specific missions. That said, in a less combustible environment, fewer peacekeepers may suffice. An approach accepting more risk might include stabilization forces in the general range of 100,000, with the U.S. contribution perhaps 30,000 to 50,000.[127]

Consider next the possibility of severe unrest in one of the world's large countries such as Indonesia, Congo, or Nigeria. At present, such problems are generally seen as of secondary strategic importance to the United States, meaning that Washington may support and help fund a peacekeeping mission under some circumstances but will rarely commit troops—and certainly will not deploy a muscular forcible intervention capability.

However, under some circumstances this situation could change. For example, if al-Qaeda developed a major stronghold in a given large country, the United States might—depending on the circumstances—consider overthrowing the country's government (if it was in cahoots with terrorists) or helping the government reclaim control over the part of its terroitory occupied by the terrorists. Or it might intervene to help one side in a civil war against another. For example, if the schism between the police and armed forces in Indonesia worsened, and one of the two institutions wound up working with an al-Qaeda offshoot, the United States might accept an invitation from the responsible half of the government to help defeat the other and the terrorist organization in question (this scenario is unlikely today, but perhaps not unthinkable for the future).[128] Or if a terrorist organization was tolerated in Indonesia, the United States might strike at it directly. That could be the case if the terrorist group took control of land near a major shipping lane in the Indonesian Straits, or simply if it decided to use part of Indonesia for sanctuary.[129]

Clearly, the requirement for foreign forces would be a function of how much of the country in question became unstable, how intact indigenous forces remained, and how large any militia or insurgent force proved to be. For illustrative purposes, if a large fraction of Indonesia or all of Congo were to become ungovernable, the problem could be twice to three times the scale of the Iraq mission. It could be five times the scale of Iraq if it involved trying to restore order throughout Nigeria, though such an operation could be so daunting that a more limited form of intervention seems more plausible—such as trying to stabilize areas where major ethnic or religious groups come into direct contact.

General guidelines for force planning for such scenarios would suggest foreign troop strength up to 100,000 to 200,000 personnel, in rough numbers, based primarily on the populations of the countries in question. That makes them not unlike the scenario of a collapsing or fracturing Pakistan. For these somewhat less urgent missions, compared to those considered in South Asia, U.S. contributions might only be 20–30 percent of the total rather than the 50 percent assumed in the preceding pages. But even so, up to two to three American divisions could be required.

QUESTION 7: Could India conquer Pakistan with conventional forces? (A case of Pakistani preemption.)

ANSWER: War between these two countries could begin over the disputed region of Kashmir, as it has several times in the past. Or a terrorist group with sanctuary in Pakistan could attack India, as one did several years ago in an assault on parliament, and as occurred again in Mumbai in late 2008. If the attack was serious enough, and was seen as benefiting from even a modest amount of active support from Pakistan's government, India might consider retaliation with an operation designed to overthrow that government or force its capitulation on terms favorable to India (perhaps involving a complete reversion of Kashmir to India). Knowing this, Pakistan might itself launch a war.

In any event, the military matchup in rough terms is as follows. Pakistan has the equivalent of roughly four ADEs in its military, along with about 100 ground attack aircraft. India has approximately eleven ADEs and about 400 ground attack planes.[130]

In broad terms, India would try to use surprise, advantages in airpower, and maneuver to overwhelm Pakistan. Given India's substantial numerical advantages, it is very hard to rule out the distinct possibility of a successful Indian attack. Knowing this, perhaps Pakistan would itself preempt India to try to benefit from surprise.

Here is how the mathematics might play out, using the method discussed earlier and based on common elements of the Kugler, Posen, and Epstein approaches. Assume that the combat exchange ratio is even—that is, Pakistan's advantages of surprise and initiative compensate for India's advantages of fighting on the defensive. This is a huge assumption and there is no way to be sure it is roughly right in advance. But as a simplifying assumption, it is not a bad place to begin. Assume further that the two air forces perform comparably, managing two sorties a day, delivering on average four munitions per sortie with a 0.1 kill probability per munition, and suffering five percent attrition per sortie due to air defenses (including ground-based systems and enemy air superiority fighters). The casualty math then looks like this, for day one of the war (further assuming a baseline of two percent attrition per day due to ground combat, and an average of 4,000 vehicles per ADE):

$$C\ (Pakistan) = (0.02) \times (4\ ADEs) = 0.08\ ADEs$$
$$C\ (India) = C\ (Pakistan) \times (exchange\ ratio)$$
$$= C\ (Pakistan) \times 1 = 0.08\ ADEs$$

The preceding is from ground combat. Then, adding in the effects of the air war:

$$C \text{ (Pakistan)} = (400 \text{ Indian ground-attack planes})$$
$$\times (2 \text{ sorties/day}) \times (4 \text{ munitions/sortie})$$
$$\times (0.1 \text{ kill probability/munition})$$
$$= 320 \text{ vehicles lost/day} = 0.08 \text{ ADEs}$$

$$C \text{ (India)} = (100 \text{ Pakistani planes}) \times (2 \text{ sorties/day}) \times (4) \times (0.1)$$
$$= 80 \text{ vehicles lost/day} = 0.02 \text{ ADEs}$$

So India's overall estimated daily losses for day one, theater-wide, would be 0.1 ADEs. Pakistan's, by contrast, would be 0.16 ADEs. Due to India's superior air force, Pakistan would lose forces faster than India would, a foreboding development for a country beginning with substantial quantitative inferiority and no hidden trump card (except, alas, its nuclear arsenal).

On subsequent days of battle, the calculations would proceed similarly, though one would first have to adjust for aircraft losses from earlier days of combat. One might also have to adjust for the activation of reserve units and their arrival at the front (though the way I have done this has assumed their mobilization, on both sides, prior to combat). Given the major advantages enjoyed by India, however, details about mobilization would matter little. Unless Pakistan could find a way to benefit from fighting on the offense more than a 1:1 exchange ratio suggests, it would likely find itself quickly in a difficult situation.

QUESTION 8: Could an airpower-based defense be used to protect a vulnerable overseas ally?

ANSWER: In the modern era, the United States has typically assumed that a country like Kuwait or Saudi Arabia would be difficult to defend from attack (from Saddam's Iraq, for example, or even from Iran). The basic assumption has been that a given sector of territory might, in effect, have to be initially conceded, then liberated later after a large buildup of some half million U.S. forces dominated by soldiers and Marines. Might it be possible instead to construct a reliable initial defense capability—even without placing huge numbers of American troops on permanent forward deployment to the region in question? If so, this could spare American allies the possible tragedy of even temporary occupation by a hostile foreign power, and could greatly reduce the expense of American military preparations for the scenario in question. This question is not particularly germane for the current Middle East, with tens of thousands of American troops in Iraq and Saddam no longer there, but it could be relevant there or elsewhere in the future.

For a modern military with good intelligence and surveillance capabilities, able to see an enemy coming, defense is—at least in theory—a more favorable form of warfare since it allows one to shoot from protected positions against exposed enemy forces. Of course, being successful on the defense is not a given historically (as noted earlier), so one must carefully assess if intelligence, surveillance, command, and control are up to the job of reliably detecting enemy movements. In addition, the defensive posture works only if a given threat is predictable enough, and a given interest important enough, to warrant deploying substantial amounts of military capability (even if not huge ground forces) in a given theater in advance of a possible crisis.

The airpower component of the Kugler, Posen, and Epstein models can be used to gain at least a very broad sense of the possibilities. Consider a situation like the defense of Saudi Arabia against Iraq. In the 1990s, several RAND analysts built on the success of airpower—and more specifically, precision munitions—in Operation Desert Storm to argue for a different approach to defend Saudi Arabia against attack. Perhaps half as much ground power as used in Desert Storm (or less) might be adequate, provided airpower was fully enabled and exploited. That would have meant purchasing and prestocking large numbers of precision munitions in the theater, ensuring that airplanes had enough refueling aircraft to deploy quickly to the theater as well as bases there from which to operate, and developing optimal reconnaissance capabilities so airpower could be directed to attack moving enemy vehicles before those vehicles could reach their originally intended destinations.[131]

The whole scheme may have assumed too much political decisiveness in Washington, Riyadh, and elsewhere about responding to concentrations of Iraqi armored power. It might also have been mistaken to imply that, with a successful defense of Kuwait or Saudi Arabia under its belt, the United States would be content to leave Saddam in power (if it wanted to overthrow him, a larger ground force would presumably have been required regardless). And the concept depended as well on prompt, unfettered access to bases that would have to be available despite the possibility of indigenous country political indecision or debilitating military attack from Saddam's forces. But the analytical framework was still quite useful for exploring options.

The RAND scholars showed what might be possible using a new type of weapon—specifically, a homing submunition that could find and attack targets largely autonomously. These submunitions, known as SKEETs, can be fired in large numbers. More than 100 can easily be carried on a single tactical combat aircraft (being launched several dozen at a time). If an enemy armored force is sufficiently concentrated, to the point where a single aircraft sortie can achieve multiple expected armor kills, then the mathematics are highly favorable to the defender—provided those munitions, and the planes that carry them, are available in adequate numbers from the opening bell of battle. Nearly 100 Iraqi armored vehicles per day were destroyed during the American air campaign in Desert Storm. It might be possible to be ten times as effective with SKEETs, even if the number of vehicles within the search zone of a given SKEET "swarm" is modest and if multiple SKEETs typically wind up striking the same target (as reflected in the denominator of the expression that follows).[132]

{[500 ground attack planes][2 sorties/day][100 SKEET/sortie][0.5 chance of finding targets/sortie][0.25 probability of kill per SKEET]}/{[10 redundant hits/average destroyed vehicle]}
= 1,250 Iraqi vehicles destroyed/day.

Of course, it is very difficult to figure out proper estimates for all the terms in the preceding simple formula. Kill probabilities of munitions can be estimated from performance on the test ranges—but data on such performances is often classified, and in any event results may be much different in a battlefield environment due to enemy countermeasures or other changed conditions. Perhaps even more crucial, and difficult to forecast accurately, is the matter of estimating the probability of detection of targets (as well as the number likely to be within range of a given aircraft all at once). Establishing a basis for determining this figure requires data from similar wars, or detailed technical information about the performance of reconnaissance assets, such as Joint STARS radar-imaging aircraft and similarly equipped unmanned aerial vehicles (UAVs). The ability to estimate such figures reliably depends in large measure on the degree to which analogous wars have been fought in recent times—or at least on the degree to which very realistic simulations and tests have been conducted. So the answer to this question is, perhaps.

QUESTION 9: Why do evenly matched militaries sometimes fight to a draw—and sometimes not?

ANSWER: The international relations literature is rife with references to balances of power. The casual suggestion of many is that countries or alliance systems that pit roughly comparable military capabilities against each other are less likely to fight, since all would know that any war would likely be long and perhaps inconclusive.

In reality, this is not true. Often, even when two sides are roughly evenly matched according to material indicators, one side wins quickly and decisively. A case in point is Germany's defeat of France in 1940, but other examples abound.

Of course, the answer has to do with military innovation and entrepreneurship, with surprise and daring, with the precepts of the Chinese tactician Sun Tze rather than the attrition-war mentality of World War I or even the American Civil War. In short, if an attacker such as Germany develops and executes a brilliant plan that its adversary had not contemplated or prepared against, it might win quickly. (An attacker might also overestimate its capabilities and suffer a terrible defeat by striking first, of course.) But how do these general, qualitative observations manifest themselves in the mathematics of the combat equations discussed earlier?

We can use either the Dupuy approach or the Posen/Kugler/Epstein concepts. Assume simply that each side has ten armored divisional equivalents or ADEs (each with an adjusted average of 25,000 troops per ADE), and 500 attack aircraft with an average payload of four munitions flying two sorties a day. This is an oversimplification, but acceptable for my illustrative purposes.

For the Dupuy model, we then need to account for the defensive advantage of the attacked country as well as the surprise advantages of the attacker. Imagine a situation like that of Germany and France in 1940, qualitatively speaking at least. Dupuy's coefficients of combat limit our choices, but we will assume what are roughly the minimum and maximum here: no advantage for the defender from its preparations, and a 2.5:1 power advantage for the attacker from surprise.[133] On top of that, either because of their superior training in general, or because of their meticulous preparation for this particular scenario, the attacking country's soldiers can be estimated to have twice the quality of the defenders at least for this scenario. All of this would yield, through the power equations:

$$P (attacker) = (250{,}000) \times (2.5) \times (2) = 1{,}250{,}000$$
$$P (defender) = 250{,}000$$

Daily casualty rates might then be, notionally:

C (defender) = (0.01) × (250,000) × (5) = 12,500 per day
C (attacker) = (0.01) × (250,000) × (1/5) = 500 per day

The terms 5 and 1/5 reflect the relative power of the two sides. The defender loss rates are in fact consistent with those of an army that collapses within a couple weeks of hard fighting, given its difficulty of moving reinforcements to (or within) the theater of combat fast enough, as well as the fact that militaries tend to collapse once they have suffered about 50 percent attrition. (To use the other models, we might assume an advantage to the attacker in the combat exchange ratio of as much as ten to one.)

QUESTION 10: How could the Persian Gulf oil economy be protected in the face of Iranian efforts to disrupt it?

ANSWER: This question resembles the complex analytical challenge of the earlier treatment of a Chinese attack on Taiwan more than the simple mathematical treatments of air-ground combat by Dupuy, Posen, Kugler, and Epstein.

In the 1980s, during the Iran–Iraq War, the United States had to address threats to shipping in the Persian Gulf. To do so, it reflagged some oil tankers under its own colors and enhanced its naval presence in the region.

This type of scenario could recur. But next time, it could do so in a more worrisome way. Given the ongoing state of serious tension in U.S.–Iranian relations, any spark could inflame a serious problem. In recent years, while Iran's arms imports have not increased as fast as some had feared, they have nonetheless permitted that country to improve its capacity to threaten shipping lanes in the narrow waters of the Persian Gulf and the Strait of Hormuz. In particular, Iran has been improving its capabilities in those very areas of military capability that could cause the United States greatest concern—advanced mines, quiet diesel submarines, and precision-guided antiship missiles.[134]

This hypothetical worry could become acute, for example, if in the coming years Israel or the United States attacked the exposed parts of Iran's nuclear infrastructure. In such an event, the United States might reinforce its defensive position in the region in advance. But an aggressive Iranian response against American friends and allies in the region, or against oil tankers in the Persian Gulf, could result anyway.

To defend the Gulf, reconnaissance and rapid-strike capabilities could be provided either via sea-based assets or land-based capabilities. Aerial and sea reconnaissance, as well as quick-strike capabilities, would be needed. American submarines would probably be desired to keep a constant track on Iranian submarines. And of course, ships to protect convoys would likely be required as well, as could mine warfare vessels.

The quantitative requirements for these various assets would be a function of three main sets of factors: geography, rotational policies, and total Iranian force strength. The United States, and any assisting allies, would need to maintain robust quick-action capabilities along the whole length of the Gulf. It would need to be able to sustain coverage twenty-four hours a day, and it would need to be able to face down an all-out Iranian assault if necessary as well.

In rough terms, these sizing criteria lead to the following rough requirements. Given Iran's small submarine force, with just three vessels, the demand for American forces would probably not require more than twenty submarines (allowing up to two U.S. subs per Iranian submarine as well as the need to rotate American ships). Even this is a very conservative estimate, since it is not apparent how functional these Iranian submarines truly are, and intelligence reports cast doubt on their likely effectiveness.

To ensure continual airspace dominance in the Gulf, roughly as many planes could be required as were needed to enforce the northern and southern no-fly zones over Iraq from 1991 to 2003—some 200 planes in all. (The size of the Persian Gulf is roughly comparable to that of Iraq, if one includes littoral regions as well as the body of water itself, so this is a reasonable approximation.) The aircraft would ideally be based at several locations along the 500 mile length of the Gulf to minimize time wasted in transit and allow for rapid reinforcement should Iran attempt an assault. In addition, some additional number of planes might need to be capable of rapid response in the face of any Iranian aerial action. (Iran's total air force numbers about 300 planes, of which perhaps 200 are airworthy.)[135]

Enough surveillance aircraft would be needed to maintain orbits at the northern and southern ends of the Gulf. Allowing for rest time for crews and maintenance time for planes, that makes for a grand total of eight to ten planes for air monitoring and a similar number for sea surveillance. (The need for two separate "orbits" of reconnaissance aircraft again accords with the Iraq no-fly-zone experience, as well as the fact that radar horizons of aircraft at 35,000 to 40,000 feet are typically about 250 miles.)[136]

In addition to its convoy escorts the United States might need to create a "fence" of ships capable of ballistic-missile defense spaced every fifty miles along Iran's seacoast to ensure short enough reaction times to any missile launch. The spacing follows from the fact that Iran's ballistic missiles would only have to travel a comparable distance, suggesting a very short flight time. Defensive interceptors would not travel notably faster than the incoming threats. There would also be a delay between the launch of the ballistic missiles and the launch of interceptor as threats were recognized and command decisions were made.

Taken together, the preceding assets resemble the air and naval components of what has commonly been considered a one-war force package in recent times. Whether some ground forces were needed as a prudent deterrent against overland Iranian aggression would also have to be considered, but the numbers here would presumably not have to reach into the "major theater war" magnitudes.

Notes

1. Geoffrey Blainey, *The Causes of War* (New York: The Free Press, 1973), pp. 246–49.

2. By "roughly comparable military capabilities," I mean a situation in which neither side is estimated to exceed the other's capabilities by more than 50 percent. See Requirements and Resources Directorate, U.S. Army Concepts Analysis Agency, *Combat History Analysis Study Effort (CHASE): Progress Report* (Washington, D.C.: Army Concepts Analysis Agency, 1986), p. 3–20, cited in Joshua M. Epstein, "Dynamic Analysis and the Conventional Balance in Europe," *International Security*, vol. 12, no. 4 (Spring 1988), p. 156.

3. Richard K. Betts, *Surprise Attack* (Washington, D.C.: Brookings, 1982), pp. 5–16.

4. Epstein, "Dynamic Analysis and the Conventional Balance in Europe," p. 156.

5. Eric V. Larson, *Casualties and Consensus* (Santa Monica, Calif.: RAND, 1996).

6. See Joshua M. Epstein, *Measuring Military Power: The Soviet Air Threat to Europe* (Princeton, N.J.: Princeton University Press, 1984), pp. xxv–xxvi.

7. For a good explanation of TACWAR, see Francis P. Hoeber, *Military Applications of Modeling: Selected Case Studies* (New York: Gordon and Breach Science Publishers, 1981), pp. 132–42.

8. Michael R. Gordon and General Bernard E. Trainor, *The Generals' War: The Inside Story of the Conflict in the Gulf* (Boston, Mass.: Little, Brown, and Co., 1995), p. 457; Department of Defense, *Conduct of the Persian Gulf War: Final Report to Congress* (Washington, D.C.: Department of Defense, April 1992), pp. A-3 through A-11; and Congressional Budget Office, "Costs of Op-

eration Desert Shield," CBO Staff Memorandum," U.S. Congress, Washington, D.C., January 1991, p. 15.

9. See, for example, John J. Mearsheimer, "Why the Soviets Can't Win Quickly in Central Europe," *International Security*, vol. 7, no. 1 (Summer 1982), reprinted in Steven E. Miller, ed., *Conventional Forces and American Defense Policy* (Princeton, N.J.: Princeton University Press, 1986), pp. 121–57; Barry R. Posen, "Measuring the European Conventional Balance: Coping with Complexity in Threat Assessment," *International Security*, vol. 9, no. 3 (Winter 1984/85), reprinted in Miller, ed., *Conventional Forces and American Defense Policy*, pp. 79–120; Joshua M. Epstein, "Dynamic Analysis and the Conventional Balance in Europe," *International Security*, vol. 12, no. 4 (Spring 1988), pp. 154–65; Eliot A. Cohen, "Toward Better Net Assessment: Rethinking the European Conventional Balance," *International Security*, vol. 13, no. 1 (Summer 1988), pp. 50–89; Steven J. Zaloga and Malcolm Chalmers, "Is There a Tank Gap?: Comparing NATO and Warsaw Pact Tank Fleets," *International Security*, vol. 13, no. 1 (Summer 1988), pp. 5–49; and Lutz Unterseher, "Correspondence: The Tank Gap Data Flap," *International Security*, vol. 13, no. 4 (Spring 1989), pp. 180–87.

10. FEBA stands for forward edge of the battle area.

11. For a good explanation and critique of the Lanchester equations, see Joshua M. Epstein, *Strategy and Force Planning: The Case of the Persian Gulf* (Washington, D.C.: Brookings Institution, 1987), pp. 146–55.

12. For an explanation of the advantages of simpler, more transparent models, see Zalmay Khalilzad and David Ochmanek, "Rethinking US Defence Planning," *Survival*, vol. 39, no. 1 (Spring 1997), pp. 43–64.

13. See Barry R. Posen, "Measuring the European Conventional Balance: Coping with Complexity in Threat Assessment," *International Security*, vol. 9, no. 3 (Winter 1984/85), reprinted in Steven E. Miller, ed., *Conventional Forces and American Defense Policy* (Princeton, N.J.: Princeton University Press, 1986), pp. 79–120.

14. Posen, "Measuring the European Conventional Balance," p. 106; and William P. Mako, *U.S. Ground Forces and the Defense of Central Europe* (Washington, D.C.: Brookings, 1983), pp. 36–37.

15. To compute ADE scores for various militaries in order to make these ADE comparisons, a system known as the WEI-WUV method is often employed. American qualitative advantages show up partly in the ADE scores, but even more in the exchange ratio. See U.S. Army Concepts Analysis Agency, War Gaming Directorate, *Weapon Effectiveness Indices/Weighted Unit Values III* (Bethesda, Md.: CAA, 1979).

It is also possible, however, to use a modified approach to scoring static inputs. Rather than employ the WEI-WUV system, a method such as TASC-FORM can be employed (the name of which derives from The Analytical Sciences Corporation, which created this database or formula). TASCFORM shows a much greater advantage for Western equipment over alternatives such as Soviet weaponry. In such a situation, if TASCFORM is used rather than WEI-WUV scoring, the exchange ratio would be less lopsided in the U.S. favor (since much of the American advantage would have already been captured in the weapons input scores).

See Lane Pierrot, *Structuring U.S. Forces After the Cold War: Costs and Effects of Increased Reliance on the Reserves* (Washington, D.C.: Congressional Budget Office, 1992), pp. 46–53; and Michael E. O'Hanlon, *The Art of War in the Age of Peace: U.S. Military Posture for the Post–Cold-War World* (Westport, Conn.: Praeger, 1992), p. 67.

16. Joshua M. Epstein, "Dynamic Analysis and the Conventional Balance in Europe," *International Security*, vol. 12, no. 4 (Spring 1988), pp. 155–58; and Joshua M. Epstein, *Conventional Force Reductions: A Dynamic Assessment* (Washington, D.C.: Brookings, 1990), pp. 51–65.

17. See, for example, Epstein, *Strategy and Force Planning*, pp. 63–88, 117–25; and Joshua M. Epstein, *Conventional Force Reductions: A Dynamic Assessment* (Washington, D.C.: Brookings, 1990), pp. 48–80.

18. It is worth noting that once breakthroughs occur, motorized armies can generally advance ten to sixty kilometers a day depending on factors like terrain and any residual resistance (in fact, even in the nineteenth century, armies sometimes averaged ten to twenty kilometers of progress a day). Against strong resistance, attackers more frequently average moving one to five kilometers a day depending on the quality and preparedness of the defense and related factors. See Trevor N. Dupuy, *Numbers, Predictions, and War: The Use of History to Evaluate and Predict the Outcome of Armed Conflict*, revised edition (Fairfax, Va.: HERO Books, 1985), pp. 16, 213–14; and Jeffrey Record, "Armored Advance Rates: A Historical Inquiry," *Military Review*, vol. 53, no. 9 (September 1973), pp. 63–66.

19. See Tami Davis Biddle, *Rhetoric and Reality in Air Warfare: The Evolution of British and American Ideas about Strategic Bombing, 1914–1945* (Princeton, N.J.: Princeton University Press, 2002), pp. 289–301.

20. Thomas A. Keaney and Eliot A. Cohen, *Gulf War Air Power Survey Summary Report* (Washington, D.C.: Government Printing Office, 1993), pp. 65, 199; and General Accounting Office (now the Government Accountability Office), *Operation Desert Storm: Evaluation of the Air Campaign* (Washington, D.C.: GAO, June 1997), GAO/NSIAD-97-134, p. 166.

21. Benjamin S. Lambeth, *NATO's Air War for Kosovo: A Strategic and Operational Assessment* (Santa Monica, Calif.: RAND, 2001), pp. 35, 62, 65; and Ivo H. Daalder and Michael E. O'Hanlon, *Winning Ugly: NATO's War to Save Kosovo* (Washington, D.C.: Brookings, 2000), pp. 135–36.

22. Trevor N. Dupuy, *Attrition: Forecasting Battle Casualties and Equipment Losses in Modern War* (Fairfax, Va.: HERO Books, 1990), p. 139.

23. Barry R. Posen, "Political Objectives and Military Options in the Persian Gulf," Defense and Arms Control Studies Working Paper, Massachusetts Institute of Technology, Cambridge, Mass. (November 1990), pp. 24–25.

24. Joshua M. Epstein, "War with Iraq: What Price Victory?" Briefing Paper, Brookings Institution, December 1990.

25. Michael R. Gordon and Bernard E. Trainor, *The Generals' War: The Inside Story of the Conflict in the Gulf* (Boston, Mass.: Little, Brown, and Co., 1995), pp. 355–80.

26. Congressional Budget Office, "Costs of Operation Desert Shield," January 1991, p. 15.

27. Dupuy, *Attrition*, pp. 73–74, 131.

28. See Directorate for Information Operations and Reports, "Persian Gulf War: Desert Shield and Desert Storm," Department of Defense, Dec. 15, 2001 (web1.whs.osd.mil/mmid/casualty); DoDefense, *Conduct of the Persian Gulf War: Final Report to Congress* (April 1992), p. M-1.

29. See also, Lawrence Freedman and Efraim Karsh, *The Gulf Conflict, 1990–1991: Diplomacy and War in the New World Order* (Princeton, N.J.: Princeton University Press, 1993), p. 409.

30. Stephen Biddle, "Victory Misunderstood: What the Gulf War Tells Us about the Future of Conflict," *International Security*, vol. 21, no. 2 (Fall 1996), pp. 139–79.

31. JSTARS was first used in Desert Storm; it can scan a region of twenty or more kilometers on a side when in broad-sweep mode. See James F. Dunnigan, *How to Make War: A Comprehensive Guide to Modern Warfare for the Post–Cold War Era* (New York: William Morrow and Company, Inc., 1993), pp. 154–55.

32. Stephen Biddle, "The Past as Prologue: Assessing Theories of Future Warfare," *Security Studies*, vol. 8, no. 1 (Autumn 1998), pp. 1–74.

33. For the sake of reference, Iraq's army of the time of roughly one million soldiers had about fifty divisions, whereas the U.S. Army of roughly 1.8 million total soldiers had only two-thirds as many, counting National Guard formations. The United States organizes a larger share of its soldiers into nondivisional support units than do many other militaries. See International Institute for Strategic Studies, *The Military Balance 1989–1990* (London: Brassey's, 1989), pp. 16–18, 101.

For reference purposes, while U.S. divisions typically have 16,000 to 18,000 soldiers, brigades have 2,000 to 5,000, battalions 250 to 1,000, companies 100 to 300, platoons 30 to 75, and squads 4 to 13. Each unit tends to have about three to five of the smaller echelon unit within it; there are three to five squads in a platoon, three to five platoons in a company, three to five companies in a battalion, five or six battalions in a brigade, and three or four brigades in a division (the terminology changes somewhat for the U.S. Marine Corps, with regiments being the closest approximation to brigades). See Robert M. Perito, ed., *Guide for Participants in Peace, Stability, and Relief Operations* (Washington, D.C.: U.S. Institute of Peace, 2007), p. 251.

Looked at another way, in the U.S. Army of 2005, there were ten active divisions and eight reserve component divisions (all the latter in the National Guard). There were also thirty-three active brigades and thirty-six reserve component brigades (most within the divisional structures). When the Army reaches its intended size of forty-three active brigades, there will also be thirty-four reserve component brigades, and if the process continues to the maximum extent now being considered, there will be forty-eight active brigades and thirty-four reserve component brigades. (Put in other terms, there were 98 active battalions and 108 reserve component battalions in 2005. The plan was to wind up with 92 active battalions and 70 reserve battalions initially, with the possibility of then going to 102 active battalions and 70 reserve battalions.) In terms of companies, the 2005 figures were 297 and 327; the interim targets are 353 and 265; the 48-brigade force plans are for 393 and 265, respectively. See Adam Talaber, *Options for Restructuring the Army* (Washington, D.C.: Congressional Budget Office, 2005), pp. 9, 15.

34. See Thomas A. Keaney and Eliot A. Cohen, *Gulf War Air Power Survey Summary Report* (Washington, D.C.: Government Printing Office, 1993), pp. 21, 58–64, 155; and Les Aspin and William Dickinson, *Defense for a New Era: Lessons of the Persian Gulf War* (Washington: Brassey's, 1992), pp. 1–41; for data on Arab–Israeli wars, see Posen, "Measuring the European Conventional Balance," in Miller, *Conventional Forces and American Defense Policy*, p. 113; and Dupuy, *Numbers, Predictions, and War*, pp. 118–39.

35. Other work, such as that done at the RAND Corporation, improves the inputs used in the air-only parts of the models. See Christopher Bowie, Fred Frostic, Kevin Lewis, John Lund, David Ochmanek, and Philip Propper, *The New Calculus* (Santa Monica, Calif.: RAND, 1993); David A. Ochmanek, Edward R. Harshberger, David E. Thaler, and Glenn A. Kent, *To Find, and Not to Yield* (Santa Monica, Calif.: RAND, 1998).

36. As noted before, it is also possible to use a modified approach to scoring static inputs. Rather than employ the WEI-WUV system, a method such as TASCFORM can be employed (whose name, again, derives from The Analytical Sciences Corporation, which created this database or formula). TASCFORM shows a much greater advantage for Western equipment over alternatives such as Soviet weaponry. It can further be used to reflect the varying degrees of skill among soldiers on various ides through its personnel weighting system. This approach is useful when trying to create a static indicator that more accurately gauges overall combat power. However, in a dynamic model, it is unnecessary, and confusing in some ways, since the issue is less about who has the better equipment and much more about who is better at employing it. (Those who doubt this might be reminded that the so-called "opposition forces" operating at America's combat training centers—American

soldiers using Russian weaponry—typically defeat the visiting main U.S. combat forces in simulated battle. In addition, as Stephen Biddle has shown for the case of Operation Desert Storm, Iraqi defeat had far more to do with poor tactics and poor use of equipment than with the quality of the equipment per se.)

See Lane Pierrot, *Structuring U.S. Forces After the Cold War: Costs and Effects of Increased Reliance on the Reserves* (Washington, D.C.: Congressional Budget Office, 1992), pp. 46–53; and Michael E. O'Hanlon, *The Art of War in the Age of Peace: U.S. Military Posture for the Post–Cold-War World* (Westport, Conn.: Praeger, 1992), p. 67.

37. Barry R. Posen, "Measuring the European Conventional Balance: Coping with Complexity in Threat Assessment," *International Security*, vol. 9, no. 3 (Winter 1984/1985), reprinted in Steven E. Miller, ed., *Conventional Forces and American Defense Policy* (Princeton, N.J.: Princeton University Press, 1986), p. 105.

38. Keaney and Cohen, *Gulf War Air Power Survey Summary Report*, pp. 105–6; and General Accounting Office, *Operation Desert Storm: Evaluation of the Air Campaign*, GAO/NSIAD-97-134, pp. 8–10, 105–7, 146–48, 157–59.

39. Civilian casualty estimates based on the briefing by William Arkin of Greenpeace to Gulf War Air Power Survey project members, Oct. 31, 1991, cited in Keaney and Cohen, *Gulf War Air Power Survey Summary Report*, p. 75; for military casualty estimates, see Keaney and Cohen, p. 107.

40. Keaney and Cohen, *Gulf War Air Power Survey Summary Report*, pp. 104–17, 203.

41. Posen, "Measuring the European Conventional Balance," p. 104.

42. Joshua M. Epstein, *Strategy and Force Planning: The Case of the Persian Gulf* (Washington, D.C.: Brookings, 1987), p. 113; and Frances M. Lussier, *Replacing and Repairing Equipment Used in Iraq and Afghanistan: The Army's Reset Program* (Washington, D.C.: Congressional Budget Office, 2007), p. xi.

43. Lane Pierrot, *Planning for Defense: Affordability and Capability of the Administration's Program* (Washington, D.C.: Congressional Budget Office, 1994), p. 22; William W. Kaufmann, *Assessing the Base Force: How Much Is Too Much?* (Washington, D.C.: Brookings, 1992), pp. 52–56; and Steven R. Bowman, "Persian Gulf War: Summary of U.S. and Non-U.S. Forces," Congressional Research Service, February 11, 1991, pp. 1–8.

44. These scores are based on the TASCFORM system of scoring more than the WEI-WUV system. See Michael E. O'Hanlon, *The Art of War in the Age of Peace* (Westport, Conn.: Praeger, 1992), p. 67.

45. Posen, "Measuring the European Conventional Balance," in Miller, ed., *Conventional Forces and American Defense Policy*, p. 133.

46. Stephen Biddle, *Military Power: Explaining Victory and Defeat in Modern Battle* (Princeton, N.J.: Princeton University Press, 2004), p. 1.

47. Historically, air-to-air combat causes substantial losses, too. As an example, the United States lost 30 percent of all aerial encounters during the

first three years of the Vietnam War. It benefited from advantages of about 10:1 in the second half of the Vietnam conflict, as it had in Korea as well as the Pacific campaign of World War II. Israel achieved aerial exchange ratios of at least 20:1 in its 1967, 1973, and 1982 engagements with Arab states. See Joshua M. Epstein, *Measuring Military Power: The Soviet Air Threat to Europe* (Princeton, N.J.: Princeton University Press, 1984), pp. 110–12. On Desert Storm, see Thomas A. Keaney and Eliot A. Cohen, *Gulf War Air Power Survey Summary Report* (Washington, D.C.: Government Printing Office, 1993), p. 61.

48. Keaney and Cohen, *Gulf War Air Power Survey*, p. 211.

49. Barry D. Watts, *The Military Uses of Space: A Diagnostic Assessment* (Washington, D.C.: Center for Strategic and Budgetary Assessments, 2001), pp. 42–3; Anne Marie Squeo, "U.S. Military's GPS Reliance Makes a Cheap, Easy Target," *The Wall Street Journal*, Sept. 24, 2002.

50. Keaney and Cohen, *Gulf War Air Power Survey Summary Report*, p. 173.

51. Biddle, "Victory Misunderstood," pp. 166–68.

52. See Trevor N. Dupuy, *Attrition: Forecasting Battle Casualties and Equipment Losses in Modern War* (Fairfax, Va.: HERO Books, 1990), pp. 104–32; see also Trevor N. Dupuy, *Numbers, Predictions, and War*, revised edition (Fairfax, Va.: HERO Books, 1985); and Trevor N. Dupuy, *Understanding War: History and Theory of Combat* (New York: Paragon Books, 1987).

53. Dupuy, *Attrition*, p. 76.

54. Ibid., *Attrition: Forecasting Battle Casualties and Equipment Losses in Modern War* (Fairfax, Va.: HERO Books, 1990), pp. 146–51.

55. ———, *If War Comes, How to Defeat Saddam Hussein* (Fairfax, Va.: HERO Books, 1991), p. 104; Congressional Budget Office, "Costs of Operation Desert Shield," Jan. 1991, p. 15.

56. See Robert L. Goldich, "Casualties and Maximum Number of Troops Deployed in Recent U.S. Military Ground Combat Actions," Congressional Research Service, Oct. 8, 1993; Brig. Gen. Robert H. Scales, *Certain Victory: The U.S. Army in the Gulf War* (Washington, D.C.: Brassey's, 1994), pp. 32–35; International Institute for Strategic Studies, *The Military Balance 1989–1990* (Washington, D.C.: Brassey's, 1989), pp. 26–27, 199; and Susan L. Marquis, *Unconventional Warfare: Rebuilding U.S. Special Operations Forces* (Washington, D.C.: Brookings, 1997), pp. 187–201.

57. See Statement of General James R. Harding, Director, Inter-American Region, Office of the Secretary of Defense, before the Subcommittee on Western Hemisphere Affairs of the House Foreign Affairs Committee, July 30, 1991 (www.nexis.com/research/search/submitViewTagged).

58. I have simplified Dupuy's method considerably. He employs factors to account for terrain, surprise, weather, and so on. More important, the way in which relative power differentials enter into his equations is not quite linear in the way I have suggested. But his method has an arbitrary quality about it at

times as well—for example, he adds a "sophistication factor" in addition to mobility, firepower, and combat effectiveness coefficients to account for the quality of one military over another. It is unclear why so many different such factors are needed to explain similar phenomena, or how one selects the proper value for each. By contrast, his methodology for computing power is relatively straightforward. For more exact information on how power ratios enter into his calculations, see Dupuy, *Attrition*, pp. 124–27, 146–52.

59. That is, 250,000 times their quality advantage of about three times a benefit from surprise of about 10 percent.

60. That is, 100,000 troops times a factor of two advantage due to the benefits of being on the defensive and fighting within a city.

61. The relative power term would not vary quite this much from one country to the other according to Dupuy's detailed tables, but for simplicity of use and for gaining an approximate sense of the calculations, these figures are not far off.

62. Dupuy, *Attrition: Forecasting Battle Casualties and Equipment Losses in Modern War* (Fairfax, Va.: HERO Books, 1990), p. 149.

63. Michael O'Hanlon, "Estimating Casualties in a War to Overthrow Saddam," *Orbis*, vol. 47, no. 1 (Winter 2003), pp. 36–37.

64. Some of my ideas here first appeared in Michael O'Hanlon, "Why China Cannot Conquer Taiwan," *International Security*, vol. 25, no. 2 (Fall 2000), pp. 51–86.

65. See U.S. Marine Corps, "Operational Maneuver from the Sea," *Marine Corps Gazette*, June 1996.

66. As noted airpower expert Richard Hallion writes, "*Air superiority* characterizes a war where a nation can exert its power over a foe with few air losses of its own, and without serious concern about the enemy's ability to contest for control of the air with its own air forces. The foe, suffering from air inferiority, can only undertake limited offensive action, and must devote the bulk of activity to defensive warfare.... *Air supremacy* implies that a nation can control a foe with essentially no or absolutely minimal air losses of its own, and without need to concern itself about the enemy's air intentions." See Richard P. Hallion, "Control of the Air: The Enduring Requirement," Air Force History and Museums Program, Bolling Air Force Base, Washington, D.C., September 1999, available at www.af.mil/shared/media/document/AFD-060726-027.pdf [accessed March 14, 2008].

67. Office of the Secretary of Defense, "Annual Report to Congress: Military Power of the People's Republic of China 2008," Department of Defense, Washington, D.C., 2008, p. 2.

68. Ministry of National Defense, Republic of China (Taiwan), *2004 National Defense Report* (Taipei, Taiwan: Ministry of National Defense, 2004), p. 29.

69. Epstein, *Measuring Military Power*, pp. 208–9, 223.

70. Personal communication from Shuhfan Ding, Institute of International Relations, National Chengchi University, Taipei, Taiwan, April 14, 2000; see also David Shambaugh, "China's Military Views the World," *International Security*, vol. 24, no. 3 (Winter 1999/2000), p. 61.

71. See, for example, David Shlapak, "Projecting Power in a China-Taiwan Contingency: Implications for USAF and USN Collaboration," in Stuart E. Johnson and Duncan Long, eds., *Coping with the Dragon: Essays on PLA Transformation and the U.S. Military* (Washington, D.C.: National Defense University, December 2007), p. 90.

The U.S. experience against Iraq in Desert Storm in 1991 provides a good window into how hard it is to shut down an enemy's air force. Coalition aircraft averaged dozens of strike sorties a day against Iraqi airfields during the war's first week, yet this did not stop the Iraqi air force from flying about forty sorties a day. That was at a time when coalition aircraft completely ruled the skies, moreover. In the airfield attacks, British planes were dropping advanced runway-penetrating weapons, precisely and from low altitude. They carried some thirty bomblets apiece, each bomblet consisting of two charges: a primary explosive to create a small hole in the runway, and a second explosive to detonate below its surface, causing a crater of ten to twenty meters' width (depending largely on soil conditions). A standard attack would have used eight aircraft, each dropping two weapons, to shut down a standard NATO-length runway of 9,000 feet by 150 feet—a difficult mission, given the need to drop the weapons at precise and quite low altitudes.

See Keaney and Cohen, *Gulf War Air Power Survey Summary Report* (Washington, D.C.: Government Printing Office, 1993), pp. 56–65; General Accounting Office, *Operation Desert Storm: Evaluation of the Air Campaign* GAO/NSIAD-97-134 (June 1997), pp. 209–12; and Christopher S. Parker, "New Weapons for Old Problems," *International Security*, vol. 23, no. 4 (Spring 1999), p. 147; Duncan Lennox, ed., *Jane's Air-Launched Weapons* (Surrey, England: Jane's Information Group, 1999), issue 33 (August 1999); and Christopher M. Centner, "Ignorance Is Risk: The Big Lesson from Desert Storm Air Base Attacks," *Airpower Journal*, vol. VI, no. 4, (Winter 1992), pp. 25–35 [available at www.airpower.maxwell.af.mil/airchronicles/apj/centner.html]; and personal communication from Dave C. Fidler, Wing Commander Air 1, British Embassy, Washington, D.C., April 14, 2000.

72. See Department of Defense, *FY04 Report to Congress on PRC Military Power: Annual Report on the Military Power of the People's Republic of China* (2004), p. 40; and International Institute for Strategic Studies, *The Military Balance 2005/2006* (Colchester: Routledge, 2005), pp. 270–76.

73. See James F. Dunnigan, *How to Make War: A Comprehensive Guide to Modern Warfare for the Post–Cold War Era*, 3rd ed. (New York: William Morrow and Company, Inc., 1993), pp. 284–92.

74. Wendell Minnick, "Washington Establishes Military Hotline with Taipei," *Jane's Defence Weekly*, October 29, 2003, p. 14.

75. Swaine, *Taiwan's National Security, Defense Policy, and Weapons Procurement Processes*, p. 60.

76. Robert Hewson, "China Boosts Its Air Assets with Ilyushin Aircraft," *Jane's Defence Weekly*, September 21, 2005, p. 16.

77. For historical perspective, see James A. Huston, "The Air Invasion of Holland," *Military Review* (September 1952), pp. 13–27; and Gerard M. Devlin, *Paratrooper!: The Saga of U.S. Army and Marine Parachute and Glider Combat Troops During World War II* (New York: St. Martin's Press, 1979).

78. This is a simplified version of Hughes' formula; see Hughes, *Fleet Tactics and Coastal Combat*, p. 268.

79. Capt. Wayne P. Hughes, Jr. (U.S. Navy, retired), *Fleet Tactics and Coastal Combat*, 2nd edition (Annapolis, Maryland: Naval Institute Press, 2000), pp. 268–79.

80. Nordeen, *Air Warfare in the Missile Age*, pp. 201–3; and *Jane's Naval Review 1987* (London: Jane's Publishing, 1987), p. 124. In flying some 300 sorties against British forces in the Falklands War, and attacking the UK ships on the open oceans where they are harder to spot than when approaching shore, Argentina sank four British ships with bombs and hit another six with bombs that did not detonate because they had been improperly fused. Argentina sank a total of six ships including those hit by Exocets and other weapons.

81. International Institute for Strategic Studies, *The Military Balance 1981/1982* (London: International Institute for Strategic Studies, 1982), pp. 92–93.

82. Ship losses are discussed in the text. As for aircraft attrition, Taiwan has well over 100 surface-to-air missile batteries with ranges of tens of kilometers—more than enough to have some coverage near all of its twenty to thirty large airfields and five major ports (the kinds of places where PRC paratroopers might do the most good, seizing assets that could then be used to deploy PRC reinforcements). In addition to its air force, it also has hundreds of anti-aircraft guns and many smaller surface-to-air missile batteries that use high-quality modified Sidewinder and Sparrow missiles.

83. Office of Technology Assessment, *Proliferation of Weapons of Mass Destruction* (Washington, D.C.: Office of Technology Assessment, 1993), pp. 45–67.

84. Dupuy, *Attrition: Forecasting Battle Casualties and Equipment Losses in Modern War* (Fairfax, Va.: HERO Books, 1990), p. 58; and Anthony H. Cordesman and Abraham R. Wagner, *Lessons of Modern War, Volume 2: The Iran–Iraq War* (Boulder, Colo.: Westview Press, 1990), p. 518.

85. See Utgoff, *The Challenge of Chemical Weapons*, pp. 148–88; and Dupuy, *Attrition*, p. 58.

86. Robert G. Nagler, *Ballistic Missile Proliferation: An Emerging Threat* (Arlington, Va.: System Planning Corporation, 1992), p. 10.

87. See Dunnigan, *How to Make War*, pp. 284–92. The typical lateral inaccuracy of gunfire or artillery fire is proportional to the distance over which the round must travel, meaning that a shot to 500 meters would be expected to have one-tenth the miss distance of a shot to five kilometers.

88. For a concurring view, see McVadon, "PRC Exercises, Doctrine, and Tactics Toward Taiwan," pp. 254–55.

89. Based on the Iran–Iraq War experience, as well as basic ballistics and blast information, warheads from a missile with several hundred kilograms of explosive might kill ten to twenty people on average, in a typical city, though the potential exists for much more devastation depending on where the explosion occurs. Artillery and mortar rounds, with warheads about one-tenth that size (say seven to twenty-five pounds of explosive, as noted here) would have about one-fifth the lethal effect. (The lethal area of an explosive varies roughly with the two-thirds power of the explosive yield. So an eightfold reduction in power produces a fourfold decrease in lethal area—and hence expected casualties.) The lethal radii for weapons in this range could be thirty to fifty meters for mortar rounds and artillery shells, and 100 meters or more for typical missile warheads. See Anthony H. Cordesman and Abraham R. Wagner, *The Lessons of Modern War, Volume II: The Iran–Iraq War* (London: Westview Press, 1990), pp. 364–68; U.S. Army, *Field Manual 5-34: Engineer Field Data* (Washington, D.C., September 1987), p. 4-1; James F. Dunnigan, *How to Make War: A Comprehensive Guide to Modern Warfare for the Post–Cold War Era* (New York: William Morrow and Company, Inc., 1993), pp. 124–25; and Janne E. Nolan, *Trappings of Power: Ballistic Missiles in the Third World* (Washington, D.C.: Brookings, 1991), pp. 68–69. I also thank Colonel Thomas Lynch of the U.S. Army for information on artillery via private correspondence, March 17, 2008.

90. John Hill, "Missile Race Heightens Tension Across Taiwan Strait," *Jane's Intelligence Review* (January 2005), pp. 44–45.

91. Anthony H. Cordesman and Abraham R. Wagner, *The Lessons of Modern War, Volume II: The Iran–Iraq War* (Boulder, Colo.: Westview Press, 1990), pp. 205–6; and Daniel L. Byman and Matthew C. Waxman, "Kosovo and the Great Air Power Debate," *International Security*, vol. 24, no. 4 (Spring 2000), pp. 37–38.

92. James C. Mulvenon, Murray Scot Tanner, Michael S. Chase, David Frelinger, David C. Gompert, Martin C. Libicki, and Kevin L. Pollpeter, *Chinese Responses to U.S. Military Transformation and Implications for the Department of Defense* (Santa Monica, Calif.: RAND Corporation, 2006), pp. 116–20.

93. John Hill, "Missile Race Heightens Tension Across Taiwan Strait," *Jane's Intelligence Review* (January 2005), pp. 44–45.

94. Bernard D. Cole, "Right-Sizing the Navy: How Much Naval Force Will Beijing Deploy?" in Roy Kamphausen and Andrew Scobell, eds., *Right-Sizing the People's Liberation Army: Exploring the Contours of China's Military*

(Carlisle, Pa.: Strategic Studies Institute, Army War College, 2007), pp. 541–42.

95. For discussion of Chinese writings that seem to take a similar tack, see Roger Cliff, Mark Burles, Michael S. Chase, Derek Eaton, and Kevin L. Pollpeter, *Entering the Dragon's Lair: Chinese Antiaccess Strategies and Their Implications for the United States* (Santa Monica, Calif.: RAND, 2007), pp. 66–73. Among other naval force modernizations, China now has about twenty modern attack submarines in its fleet, and it is also expected to acquire ocean reconnaissance satellites (early versions of which it already reportedly possesses) as well as communications systems capable of reaching deployed forces in the field in the next five to ten years. See Office of the Secretary of Defense, *Military Power of the People's Republic of China, 2008: Annual Report to Congress* (Washington, D.C.: Department of Defense, 2008), available at www.defenselink.mil/pubs/pdfs/China_Military_Report_08.pdf, pp. 4, 27 [accessed March 20, 2008]; and Michael McDevitt, "The Strategic and Operational Context Driving PLA Navy Building," in Roy Kamphausen and Andrew Scobell, eds., *Right-Sizing the People's Liberation Army: Exploring the Contours of China's Military* (Carlisle, Pa.: Strategic Studies Institute, Army War College, 2007), p. 499.

96. Capt. Wayne P. Hughes, Jr. (U.S. Navy, retired), *Fleet Tactics and Coastal Combat*, 2nd edition (Annapolis, Maryland: Naval Institute Press, 2000), pp. 224–27.

97. O'Hanlon, *Defense Policy Choices for the Bush Administration*, p. 189.

98. See Congressional Budget Office, *U.S. Naval Forces: The Sea Control Mission* (Washington, D.C.: Congressional Budget Office, 1978).

99. For analyses that remain mostly correct today, see Owen Cote, *The Future of Naval Aviation* (Cambridge, Mass.: MIT Security Studies Program, 2006), pp. 34–37, available at http://web.mit.edu/ssp/; Owen Cote and Harvey Sapolsky, *Antisubmarine Warfare After the Cold War* (Cambridge, Mass.: MIT Security Studies Program, 1997), p. 13; and Tom Stefanick, *Strategic Antisubmarine Warfare and Naval Strategy* (Lexington, Mass.: Lexington Books, 1987), pp. 35–49.

100. Capt. Wayne P. Hughes, Jr. (U.S. Navy, retired), *Fleet Tactics and Coastal Combat*, 2nd edition (Annapolis, Maryland: Naval Institute Press, 2000), pp. 172–73.

101. Desmond Ball, "China Pursues Space-Based Intelligence Gathering Capabilities," *Jane's Intelligence Review* (December 2003), pp. 36–39.

102. Michael E. O'Hanlon, *Neither Star Wars Nor Sanctuary: Constraining the Military Uses of Space* (Brookings, 2004), pp. 91–104; Bill Gertz, "Chinese Missile Has Twice the Range U.S. Anticipated," *The Washington Times*, November 20, 2002, p. 3; Barry Watts, *The Military Uses of Space: A Diagnostic Assessment* (Washington, D.C.: Center for Strategic and Budgetary Assessments, 2001); Bob Preston, Dana J. Johnson, Sean J. A. Edwards, Michael

Miller, and Calvin Shipbaugh, *Space Weapons, Earth Wars* (Santa Monica, Calif.: RAND, 2002); and Benjamin Lambeth, *Mastering the Ultimate High Ground: Next Steps in the Military Uses of Space* (Santa Monica, Calif.: RAND, 2003).

103. For one good concise history of early targeting and force planning ideas, see Desmond Ball, "The Development of the SIOP, 1960–1983," in Desmond Ball and Jeffrey Richelson, *Strategic Nuclear Targeting* (Ithaca, N.Y.: Cornell University Press, 1986), pp. 57–83.

104. See Bruce G. Blair, *Strategic Command and Control: Redefining the Nuclear Threat* (Washington, D.C.: Brookings, 1985); and Bruce G. Blair, *Global Zero Alert for Nuclear Forces* (Washington, D.C.: Brookings, 1995).

105. See Scott D. Sagan, *The Limits of Safety: Organizations, Accidents, and Nuclear Weapons* (Princeton, N.J.: Princeton University Press, 1993).

106. See Barry R. Posen, *Inadvertent Escalation: Conventional War and Nuclear Risks* (Ithaca, N.Y.: Cornell University Press, 1991).

107. Matthew Bunn and Kosta Tsipis, "The Uncertainties of a Preemptive Nuclear Attack," *Scientific American* (November 1983).

108. David Mosher, "Appendix B: Exchange Calculations," in Raymond Hall, David Mosher, and Michael O'Hanlon, *The START Treaty and Beyond* (Washington, D.C.: Congressional Budget Office, 1991), pp. 143–65. Other examples include the Soviet SS-19 with a lethal radius of 300 meters, the U.S. Minuteman IIIA with a lethal radius of 185 meters, and the U.S. D5/Mark 5 (Trident II SLBM) warhead with a lethal radius of 210 meters.

The lethal radius of a warhead is calculated by taking the one-third power of (Y/H), where Y is the attacking warhead yield measured in megatons and H the hardness of the attacked missile silo in thousands of pounds per square inch, and then multiplying that by 460 meters. So a two megaton warhead attacking a 2,000 pound/square inch silo would have a lethal radius of 460 meters.

The single-shot survival probability for a silo attacked by a given warhead is then calculated as follows. First, square the lethal radius of the warhead (that is, take it to the second power, or multiply it by itself). Then, square the circular error probable of the warhead, and multiply the resulting number by 1.44, and then make it negative. Take the ratio of the first term (the square of the lethal radius) to the second (the negative of 1.44 times the square of the circular error probable). That overall expression is then the power to which the naturally occurring number "e" is taken. In short, Single Shot Survival = [exp][-LRsquared/1.44CEPsquared], where LR = lethal radius and CEP = circular error probable.

109. Samuel Glasstone, ed., *The Effects of Nuclear Weapons*, revised edition (Washington, D.C.: Government Printing Office, 1962), pp. 134–35, 156–76; Mosher, "Appendix B: Exchange Calculations," in Hall, Mosher, and O'Hanlon, *The START Treaty and Beyond*, p. 159; and David Ochmanek and

Lowell H. Schwartz, *The Challenge of Nuclear-Armed Regional Adversaries* (Santa Monica, Calif.: RAND, 2008), p. 7. Bombers would likely be damaged once struck by two to four pounds per square inch of overpressure or more.

110. See Sumit Ganguly, *Conflict Unending: India–Pakistan Tensions Since 1947* (New York: Columbia University Press, 2001).

111. See Stephen Philip Cohen, *The Idea of Pakistan* (Washington, D.C.: Brookings, 2004), pp. 97–130.

112. See International Crisis Group, *Unfulfilled Promises: Pakistan's Failure to Tackle Extremism* (Brussels, 2004).

113. General David H. Petraeus, Lt. General James F. Amos, and Lt. Colonel John A. Nagl, *The U.S. Army/Marine Corps Counterinsurgency Field Manual* (Chicago: University of Chicago Press, 2007), p. 23.

114. James Dobbins, John G. McGinn, Keith Crane, Seth G. Jones, Rollie Lal, Andrew Rathmell, Rachel Swanger, and Anga Timilsina, *America's Role in Nation Building from Germany to Iraq* (Santa Monica, Calif.: RAND, 2003), p. xvii.

115. Michael E. O'Hanlon, *Saving Lives with Force* (Washington, D.C.: Brookings, 1997), pp. 38–42; and James T. Quinlivan, "Force Requirements in Stability Operations," *Parameters*, vol. 25, n. 4 (Winter 1995–1996), pp. 59–69.

116. Seth G. Jones, Jeremy M. Wilson, Andrew Rathmell, and K. Jack Riley, *Establishing Law and Order After Conflict* (Santa Monica, Calif.: RAND, 2005), p. 19; James Dobbins, John G. McGinn, Keith Crane, Seth G. Jones, Rollie Lal, Andrew Rathmell, Rachel Swanger, and Anga Timilsina, *America's Role in Nation-Building: From Germany to Iraq* (Santa Monica, Calif.: RAND, 2003); and James Dobbins, Seth G. Jones, Keith Crane, Andrew Rathmell, Brett Steele, Richard Teltschik, and Anga Timilsina, *The UN's Role in Nation-Building: From the Congo to Iraq* (Santa Monica, Calif.: RAND, 2005).

117. See Center on International Cooperation, *Annual Review of Global Peace Operations, 2007* (Boulder, Colo.: Lynne Rienner Publishers, 2007), p. 3; International Institute for Strategic Studies, *The Military Balance 2008* (Oxfordshire, England: Routledge, 2008), p. 448, plus enclosed map; and Michael E. O'Hanlon, *Expanding Global Military Capacity for Humanitarian Intervention* (Washington, D.C.: Brookings, 2003), pp. 56–57.

118. International Institute for Strategic Studies, *The Military Balance 2003–2004*, pp. 140–42.

119. Angel Rabasa, Lesley Anne Warner, Peter Chalk, Ivan Khilko, and Paraag Shukla, *Money in the Bank: Lessons Learned from Past Counterinsurgency (COIN) Operations* (Santa Monica, Calif.: RAND, 2007), pp. ix–xv, 1–4.

120. Andrew F. Krepinevich, Jr., *The Army and Vietnam* (Baltimore, Md.: Johns Hopkins Press, 1986), pp. 177–214; and Robert S. McNamara, *In*

Retrospect: The Tragedy and Lessons of Vietnam (New York: Vintage Books, 1995), pp. 169–77, 210–12, 220–23, 233–47, 262–63, 282–93.

121. Neil Sheehan, *A Bright Shining Lie* (New York: Vintage Books, 1988), pp. 201–65.

122. David Galula, *Counterinsurgency Warfare: Theory and Practice* (New York: Praeger, 2005), pp. 70–86.

123. Krepinevich, *The Army and* Vietnam, pp. 172–77.

124. General David H. Petraeus, Lt. General James F. Amos, and Lt. Colonel John A. Nagl, *The U.S. Army/Marine Corps Counterinsurgency Field Manual* (Chicago: University of Chicago Press, 2007), pp. 1–52; Steven Metz, *Learning from Iraq: Counterinsurgency in American Strategy* (Carlisle, Pa.: Army War College Strategic Studies Institute, 2007), pp. 1–30; and Thomas E. Ricks, *Fiasco: The American Military Adventure in Iraq* (New York: Penguin Press, 2006), pp. 149–202.

125. International Institute for Strategic Studies, *The Military Balance 2003–2004*, pp. 136–37, 337.

126. General David H. Petraeus, Lt. General James F. Amos, and Lt. Colonel John A. Nagl, *The U.S. Army/Marine Corps Counterinsurgency Field Manual* (Chicago: University of Chicago Press, 2007), p. 23.

127. Michael E. O'Hanlon, *Saving Lives with Force* (Washington, D.C.: Brookings, 1997), pp. 38–42; and James T. Quinlivan, "Force Requirements in Stability Operations," *Parameters*, vol. 25, n. 4 (Winter 1995–1996), pp. 59–69.

128. On Indonesia, see Robert Karniol, "Country Briefing: Indonesia," *Jane's Defence Weekly*, April 7, 2004, pp. 47–52.

129. Krepinevich, *The Conflict Environment of 2016*, pp. 23–27.

130. In theory, calculating ADE totals from equipment inventories and force structure tables (as listed in available databases) is quite difficult and tedious. In practice, a good approximation can be obtained by beginning with active equipment inventories, starting with main battle tanks. For example, India has about 4,000 in its active stocks, featuring 330 T-90 and almost 2,000 T-72 and about 1,000 indigenous Vijayanta tanks (as well as some 700 T-55s being gradually replaced by T-90s). Typically, there are 300 main battle tanks in a heavy division, so these inventories are enough for thirteen divisions in principle—though perhaps just eleven to twelve in practice assuming a modest reserve of excess tanks to replace those in depot or otherwise temporarily unserviceable. All of these tanks except the T-55 are reasonably modern and within 10 to 20 percent of the "score" of an M1 tank (following the conservative methodology of the WEI-WUV system). The T-55 stocks by contrast are much inferior, worth at most half what a modern tank might be in terms of quality. So compensating for this fact, and the attrition/maintenance reserve mentioned here, India might have the equivalent of ten ADEs of tank inventory. See International Institute for Strategic Studies, *The Military Balance 2008* (Oxfordshire, England: Routledge, 2008), pp. 341–51.

131. Christopher Bowie, Fred Frostic, Kevin Lewis, John Lund, David Ochmanek, and Philip Propper, *The New Calculus: Analyzing Airpower's Changing Role in Joint Theater Campaigns* (Santa Monica, Calif.: RAND, 1993), pp. xi–xxii.

132. See David A. Ochmanek, Edward R. Harshberger, David E. Thaler, and Glenn A. Kent, *To Find, and Not to Yield: How Advances in Information and Firepower Can Transform Theater Warfare* (Santa Barbara, Calif.: RAND, 1998), pp. 6–12, 31–42, 77–85.

133. Trevor N. Dupuy, *Attrition: Forecasting Battle Casualties and Equipment Losses in Modern War* (Fairfax, Va.: HERO Books, 1990), p. 151.

134. Krepinevich, *The Conflict Environment of 2016: A Scenario-Based Approach* (Washington, D.C.: Center for Strategic and Budgetary Assessments, 1996), pp. 11–15; and Caitlin Talmadge, "Closing Time: Assessing the Iranian Threat to the Strait of Hormuz," *International Security*, vol. 33, no. 1 (Summer 2008), pp. 82–117.

135. International Institute for Strategic Studies, *The Military Balance 2008*, p. 243.

136. The formula for radar horizon is RH=square root of (diameter of Earth×altitude of satellite). This formula follows directly from the Pythagorean theorem, drawing a right triangle with one side the radius of the Earth, a second side the distance from the satellite in question to the farthest point on Earth's surface within its view, and a third side from the center of the Earth to the satellite (this latter segment is the triangle's hypotenuse). Symbolically, $RH = \sqrt{(DA)}$. Since the diameter of the Earth is about 8,000 miles, an aircraft at just under eight miles' altitude can see about 250 miles.

Key References and Suggestions for Further Reading

Betts, Richard K., *Surprise Attack* (Washington, D.C.: Brookings, 1982).

Biddle, Stephen, *Military Power: Explaining Victory and Defeat in Modern Battle* (Princeton, N.J.: Princeton University Press, 2004).

Blair, Bruce G., *Strategic Command and Control: Redefining the Nuclear Threat* (Washington, D.C.: Brookings, 1985).

Bowie, Christopher, Fred Frostic, Kevin Lewis, John Lund, David Ochmanek, and Philip Propper, *The New Calculus* (Santa Monica, Calif.: RAND, 1993).

Bunn, Matthew, and Kosta Tsipis, "The Uncertainties of a Preemptive Nuclear Attack," *Scientific American* (November 1983).

Center on International Cooperation, *Annual Review of Global Peace Operations, 2007* (Boulder, Colo.: Lynne Rienner Publishers, 2007).

Cordesman, Anthony H., and Abraham R. Wagner, *Lessons of Modern War, Volume 2: The Iran–Iraq War* (Boulder, Colo.: Westview Press, 1990).

Cote, Owen and Harvey Sapolsky, *Antisubmarine Warfare After the Cold War* (Cambridge, Mass.: MIT Security Studies Program, 1997).

Dobbins, James, John G. McGinn, Keith Crane, Seth G. Jones, Rollie Lal, Andrew Rathmell, Rachel Swanger, and Anga Timilsina, *America's Role in Nation Building from Germany to Iraq* (Santa Monica, Calif.: RAND, 2003).

Dunnigan, James F., *How to Make War: A Comprehensive Guide to Modern Warfare for the Post–Cold War Era* (New York: William Morrow and Company, Inc., 1993).

Dupuy, Trevor N., *Numbers, Predictions, and War: The Use of History to Evaluate and Predict the Outcome of Armed Conflict*, revised edition (Fairfax, Va.: HERO Books, 1985).

Epstein, Joshua M., *Strategy and Force Planning: The Case of the Persian Gulf* (Washington, D.C.: Brookings Institution, 1987).

General Accounting Office (now the Government Accountability Office), *Operation Desert Storm: Evaluation of the Air Campaign* (Washington, D.C.: GAO, June 1997), GAO/NSIAD-97-134.

Hoeber, Francis P., *Military Applications of Modeling: Selected Case Studies* (New York: Gordon and Breach Science Publishers, 1981).

Hughes, Capt. Wayne P., Jr. (U.S. Navy, retired), *Fleet Tactics and Coastal Combat*, 2nd edition (Annapolis, Maryland: Naval Institute Press, 2000).

Johnson, Stuart E. and Duncan Long, eds., *Coping with the Dragon: Essays on PLA Transformation and the U.S. Military* (Washington, D.C.: National Defense University, December 2007).

Kamphausen, Roy and Andrew Scobell, eds., *Right-Sizing the People's Liberation Army: Exploring the Contours of China's Military* (Carlisle, Pa.: Strategic Studies Institute, Army War College, 2007).

Keaney, Thomas A. and Eliot A. Cohen, *Gulf War Air Power Survey Summary Report* (Washington, D.C.: Government Printing Office, 1993).

Krepinevich, Andrew F., *The Conflict Environment of 2016* (Washington, D.C.: Center for Strategic and Budgetary Assessments, 1996).

Lambeth, Benjamin S., *NATO's Air War for Kosovo: A Strategic and Operational Assessment* (Santa Monica, Calif.: RAND, 2001).

Miller, Steven E., ed., *Conventional Forces and American Defense Policy* (Princeton, N.J.: Princeton University Press, 1986).

Mulvenon, James C., Murray Scot Tanner, Michael S. Chase, David Frelinger, David C. Gompert, Martin C. Libicki, and Kevin L. Pollpeter, *Chinese Responses to U.S. Military Transformation and Implications for the Department of Defense* (Santa Monica, Calif.: RAND Corporation, 2006).

Petraeus, General David H., Lt. General James F. Amos, and Lt. Colonel John A. Nagl, *The U.S. Army/Marine Corps Counterinsurgency Field Manual* (Chicago: University of Chicago Press, 2007).

Posen, Barry R., *Inadvertent Escalation: Conventional War and Nuclear Risks* (Ithaca, N.Y.: Cornell University Press, 1991).

CHAPTER III

Logistics and Overseas Bases

THE FAMOUS ADAGE that while civilians think strategy, generals think logistics is only a slight exaggeration. Without transportation, bases from which to operate, and support units, combat units are of little inherent use unless a battle comes straight to them and they can fight on their home fields, so to speak. Even then, tactical mobility requires strong attention to logistics.

For some observers, what separates the U.S. armed forces from all others is the large Pentagon budget, making it possible to buy space assets, advanced fighters and submarines, precision munitions, a large nuclear arsenal, missile defenses, and the like. All that is significant, to be sure. But it is arguably the case that what most separates the U.S. military from all others is its ability to deploy, operate, and sustain itself abroad indefinitely. Supply trucks, mobile depots, and ferry-like transport ships are just as important as stealth bombers and laser-guided or satellite-guided bombs. Almost no other country in the world can conduct major operations anywhere except on home territory.

The key elements of a logistics capability permitting operations abroad are threefold. First, the transportation assets to move combat forces promptly and efficiently to where they are needed—ideally even if infrastructure is lacking or damaged at the destination. This capability is needed even when forces are permanently deployed abroad, since most major operations require reinforcements from stateside locations. Second, the base network needed to carry out this deployment and to receive forces once they are deployed. Some of this can be constructed as needed, but that is a slow process. Third, support assets to provide the fuel, food, water, ammunition, equipment repair capabilities, medical care, and other materials or services required by a military at war. And of course, these support assets need to be transported somehow, and based somewhere, themselves. They also should be capable of autonomous operations without the help of indigenous civilian or military infrastructure and personnel. Increasingly, though, they may involve a large role for civilian contractors acting in a quasi-military capacity—a trend in need of greater oversight, to be sure, but one that is probably irreversible as well.

Understanding logistics is crucial to successful military operations. Alas, it is often confusing. The issue is less the conceptual complexity of the subject and more the sheer amount of detail—and huge amount of material—involved in moving large forces. One can get lost in data unless one has a few clear organizing principles, frames of reference, and rules of thumb to provide broad structure to the subject.

To give one example about how difficult logistics can be, the debate about which U.S. forces to send to Central Command, and when, and in what order preoccupied the Pentagon in the winter of 2003 just prior to the invasion of Iraq. Secretary Rumsfeld's desire to minimize the U.S. footprint and his insistence on continually reexamining the Time-Phased Force Deployment List (or TPFDL, a detailed schedule governing the military's transport of combat units to the vicinity of Iraq) literally consumed weeks of time for thousands of people.

Getting all the details of actual deployments right—and conducting them correctly—requires lots of computerization, the use of barcodes on supplies, standardization of equipment, and sheer practice. But understanding the basic constraints on what is doable and what is not is less difficult. Fairly simple rules of thumb and careful use of arithmetic can aid in understanding key policy questions, such as how much dedicated airlift and sealift to purchase, which overseas military bases are most important, and which responsibilities to outsource to private contractors.

This chapter is the shortest of the book because it is on a fairly specific topic. But that topic merits a full chapter given its centrality in defense planning for all who do it seriously, and its crucial character in war planning for any military that would like to prevail in combat. It also fits very naturally into a book on defense analysis because many if not most aspects of logistics lend themselves to quantification and structured analysis—with the caveat that like most things in warfare, even the best laid logistics plans and operations never survive contact with the enemy (or even, quite often, contact with natural factors like the weather).

The chapter divides logistics issues into two broad categories: transportation and bases. The former involves moving not only people and combat equipment, but supplies. Moreover, there are several key stages in movement, including the intercontinental aspect of transport and the local matter of keeping units provisioned as they conduct operations and move about. The subject of bases covers airfields, ports, troop barracks and training ranges, supply and maintenance depots, reconnaissance and communications facilities, and the like.

The principal focus here, as in other parts of the book, is on the U.S. military. However, much of the discussion can be easily generalized.

Strategic and Tactical Transportation

How does one move the equivalent of a mid-sized city halfway across the world, and do so quickly and with enough care and prudence to avoid possible enemy attack in the process? Then, how does one keep that city functioning as it spreads out on a battlefield and engages with the enemy? These are the typical logistical challenges faced by the United States when conducting combat planning.

The General Challenge of Overseas Military Operations

Other countries sometimes try to move several thousand troops within a period of months—and often face problems in doing so. By contrast, the United States plans to move a quarter million to a half million troops and their associated equipment as well as their human necessities within a similar period of several months. In fact, it has done so twice since the Cold War ended, both times in preparation for war against Iraq. It continues planning so as to be able to do so again if need be in places such as Korea.

The fact that military transportation is difficult should be obvious from the broad numbers—preparing for a major war overseas may not involve building skyscrapers or schools or factories, but it does require relocating most other elements of the equivalent of a mid-sized city like Washington, D.C. (population 550,000). In Operation Desert Storm, the United States moved not only half a million people, and more than 100,000 vehicles, but a total of about 10 million tons of supplies (more than 6 million of which were petroleum products). The numbers would have been even larger absent Saudi host nation support—each day the Saudis provided a quarter million meals and two million gallons of potable water (after four months, that would add up to another million tons of supplies).[1] Regional partners also provided up to 15 million gallons of fuel per day for the Air Force alone (about 50,000 tons).[2] Indeed, strategic transportation is so difficult that the United States itself, if caught off guard, can have difficulty with even modest movements of troops and equipment—as when it took a month to move twenty-four artillery launchers and backup capabilities into the Balkans during the Kosovo War of 1999.

TABLE 3.1
Basic Requirements for Desert Storm

500,000 troops

More than 10 million total tons of supplies

Three million tons of bulk supplies (ammunition, spare parts, vehicles, etc.)

More than 6 million tons of fuel shipped from the U.S., plus 2 to 3 million tons more provided by local partners

One million tons of water

This chapter does not provide an actual deployment plan for any concrete operation, of course. Doing so requires detailed examination of which types of ships and planes are available for a given contingency, what supplies and equipment each can hold and in what quantity—not to mention detailed discussion of where supplies will be loaded and unloaded, and what equipment is available to do the loading and unloading. The main purpose here is to provide orders of magnitude. These are useful for estimating the costs of various transportation capabilities, for understanding how to move huge amounts of tonnage over large distances quickly, and for assessing the relative importance of different foreign military bases.

In laying out the basics of strategic transportation, it is useful to focus on a few key numbers. Table 3.1 shows some of the key numbers for Operation Desert Storm.

Specific Rules of Thumb on Transport and Logistics

So much for the broad overview and big numbers. For considering other possible operations, including smaller ones, it is necessary to be able to break down some of these big numbers unit by unit. For example, the key Army formations shown in Table 3.2 have the following rough weights (this data comes from the late 1990s, but for main combat formations, there has been only modest change since):[3]

Army formations typically consume the amounts of supplies shown in Table 3.3 (Marine Corps units of comparable size, being intermediate in their weight, typically consume quantities of supplies falling between the Army light division and heavy division estimates):

TABLE 3.2
Tonnage of Standard American Army Divisions

17,000 tons for a light division (11,000 troops)

27,000 tons for the 82nd Airborne Division (just over 13,000 people)

36,000 tons for the 101st Air Assault Division (nearly 16,000 soldiers)

110,000 tons for an armored or mechanized division (18,000 soldiers)

99,000 tons for an Army Corps structure, for the command and coordination and general support of several divisions in the field (22,000 troops)

TABLE 3.3
Logistics Supply Requirements for Forces in Combat

---

600 tons per day for a heavy Army brigade of 3,100 soldiers in combat (roughly 300 tons of fuel, 130 tons of water, 85 tons of dry stores and 60 tons of ammunition)

300 tons per day for an airborne brigade of about 3,400 soldiers in combat (roughly 85 tons of fuel, 50 tons of dry stores, 145 tons of water, 10 tons of ammunition)

400 tons per day for a Stryker medium-weight brigade of about 3,900 soldiers (110 tons of fuel, 170 tons of water, 70 tons of dry stores, 40 tons of ammunition)*

Source: Robert W. Button, John Gordon IV, Jessie Riposo, Irv Blickstein, and Peter A. Wilson, *Warfighting and Logistic Support of Joint Forces from the Joint Sea Base* (Santa Monica, Calif.: RAND, 2007), pp. 77–85.

*The main equipment in a heavy brigade includes 58 M1A2 tanks, 109 Bradley Fighting Vehicles, 43 armored personnel carriers, 45 HMMWVs, 23 recovery vehicles, 451 utility trucks, and 218 cargo trucks. The main equipment in an infantry brigade includes 75 HMMWVs, 16 heavy trucks, 25 medium tactical vehicles, 13 light tactical vehicles, and 263 utility trucks. The main equipment in a Stryker brigade includes 302 Stryker vehicles, 381 utility trucks, and 158 cargo trucks.

A division typically has three or four main brigades plus various supporting capabilities. So to calculate the daily tonnage requirements for a heavy division, one could start with the brigades: 600 tons per day times three to four brigades per division. There would be another 6,000 to 8,000 soldiers in such a division, not tied to brigades, who might consume roughly half as many supplies per person as those soldiers within brigades proper. So the entire division might use 3,000 tons a day of supplies, in rough numbers.

In rare instances, units in the field can be supplied directly from the United States. However, it is much more common that they require tactical-level resupply operations to reach them wherever they have moved on the battlefield. In other words, the United States would use large intercontinental transport assets to create major rear-area logistics bases in a theater of operations, and then use those main bases to supply forces in the field. Aircraft can help with this job, generally in the form of C-130 propeller aircraft and transport helicopters, but large operations require overland resupply. Most commonly that means trucks. And lots of them.

For the United States in recent operations, moving supplies over desert in situations where it controlled the air, theater resupply operations have seemed relatively straightforward. But this is not always the case. A brilliant study by Joshua Epstein in the mid-1980s, at a time when the Soviet Union still occupied Afghanistan and was feared by some to have designs on Iran as well, underscored the difficulty of supplying large armies over great distances by truck in the face of enemy opposition and in the context of complex topography. Because the region between the Soviet Union and central as well as coastal Iran passed through mountains, the Soviets could not have carried out off-road movements very effectively. They would have been constrained to use a modest number of roads, subject to sabotage and both land and aerial attack, in places that would have had to handle a very large number of vehicles per day. They would also have likely been constrained by the sheer availability of trucks, which not all armies maintain in adequate numbers. This is an example worth bearing in mind for the future, even if battles against Soviet forces in the interior of Asia are no longer so plausible.

As noted, most brigades require several hundred tons of supplies a day, and trucks typically have payloads of five to ten tons. So roughly 100 trucks per day are needed to supply a given brigade in combat, especially once the brigade's typical support units are also accounted for. A large military operation involving twenty brigades, plus associated support elements, would therefore require 1,000 to 4,000 truckloads of logistics support a day in round numbers, depending on the amount of combat and the amount of movement the typical unit is conducting. If these supplies need to move on roads, given the topography in question, bottlenecks can develop. A single major artery might be able to handle 2,000 truckloads of supplies a day—but only if it is possible to send out a fresh armed convoy including twenty supply trucks every fifteen minutes and maintain that pace continuously. The United States has enough trucks to maintain such a pace in its inventories.[4] Still, this scale of resupply is generally only feasible if vehicles are in very good shape (since frequent breakdowns will interfere with movement), if roads do not need to be shared with civilian traffic, if key infrastructure like bridge networks is robust (and defensible), if weather is not a major factor, and if engineering and maintenance crews and equipment can keep the roads in acceptable driving condition.

For movement of Air Force units, the lift problem is different because most of the combat platforms themselves are naturally self-deploying. They generally require refueling, in the air or on land bases or both, but one does not focus on the weight of the combat systems when doing the arithmetic of strategic transport. This is not to say, however, that moving air combat wings is trivial. Three types of supplies—fuel, munitions, and base support assets such as fuel distribution systems, aircraft maintenance equipment, runway maintenance equipment, missile defenses, and command and control—can be quite heavy. Spare parts impose additional demands.

In Operation Desert Storm, the United States averaged seventy intercontinental flights to the theater a day. Its average rate of delivering tonnage was 1,700 tons per day for the first month of the deployment and 3,600 tons per day in January of 1991, the peak month.[5] At that latter rate, it would take about two weeks to deploy the needed support for 800 aircraft—assuming that 25 percent of the lift is devoted to the Air Force, that some munitions are eventually provided by prepositioned ships (and later by sealift), and that a substantial amount of fuel is either predeployed or provided by a regional partner such as Saudi Arabia.[6] Putting this in terms of the support and supply requirements for a combat wing of seventy-two fighter aircraft (about 1,500 to 2,000 people) gives the following rough numbers shown in Table 3.4.

TABLE 3.4
Key Tonnage Data for a Typical Air Force Combat Wing (72 aircraft)

1,000 to 2,000 tons of ground-support equipment

100 to 200 tons per day of ammunition expenditure

500,000 to 1,000,000 gallons of fuel, or up to 3,500 tons a day

What is needed to move all these people and their supplies? Obviously, the two main ways to move intercontinentally are airlift and sealift. Airlift can be subdivided into categories: dedicated military lift for cargo, largely commercial aircraft for people, and tankers for (limited amounts of) fuel. The sealift can be subdivided into two main categories: ships for moving equipment and ships for moving petroleum. The former type of sealift can be further broken down into ships readily available and easy to load ("roll on / roll off ships"), as well as ships that need to be taken out of "mothballs" or rented from the commercial sector (the latter typically require cranes for loading and unloading in port).

Airlift is employed to move people certainly, but it is also frequently used to transport sensitive electronics, helicopters, high-priced and scarce ammunition, and critical spare parts. It is sometimes used for equipment needed promptly, like recently manufactured Mine Resistant Ambush Protected (MRAP) vehicles in Iraq and Afghanistan—several thousand of which were airlifted to the Central Command theater in recent years. (Arguably, not all needed to be airlifted, but there was time pressure to get at least the first batches to the theater as fast as possible.)[7] Sealift is used to move fuel, water, vehicles, bulk ammunition, makeshift housing, mobile depots and hospitals, and in general the heaviest materials.

The key factors for understanding the respective transport capacities of planes and ships are their speeds, their payloads, their dependability/availability, and their time needed for loading as well as unloading. For all major U.S. planes, average speeds are about 500 miles per hour and average loading and unloading times about three to four hours at each end of operations (with the C-17 slightly faster than others). Average payloads and average aircraft utilization rates per day are shown in Table 3.5.

TABLE 3.5
Data on Transport Aircraft

---

23 tons and about ten hours for the (now retired) C-141

61 tons and 8.1 hours for the C-5

45 tons and 12.5 hours for the C-17

33 tons and 8.6 hours for the KC-10

Source: David Arthur, *Options for Strategic Military Transportation Systems* (Washington, D.C.: Congressional Budget Office, September 2005), pp. 8, 14; Schmidt, *Moving U.S. Forces*, p. 13.

For shorter-range transports such as the V-22 Osprey tilt-rotor aircraft and CH-53 helicopter, typical payloads are about ten tons.[8] The payload of the C-130 propeller aircraft, the workhorse of intra-theater airlift for the U.S. military, is about fifteen tons and its range varies from 1,200 to 2,000 miles at normal payload depending on the version of the aircraft at issue.[9]

Aircraft range depends on how much cargo is carried. Generally speaking, it is not a matter of cargo physically displacing fuel, in the sense of taking up space that could hold extra fuel canisters, but rather of the aircraft's maximum takeoff weight requiring a tradeoff between the two. If the cargo in question is made up of bulky, relatively light matériel, there may be no real tradeoff; it might be possible to fill up fuel bays and cargo holds entirely. (At some level, adding any weight to an airplane reduces its fuel efficiency in flight; airplanes are somewhat more efficient when lighter. But this consideration is typically minor; aircraft fuel efficiencies typically vary by 5 to 10 percent depending on weight.) Quite often, however, equipment is heavy enough that a full load of cargo requires some reduction in fuel, so that the airplane stays within maximum takeoff weight restrictions.

As one example, consider these facts and figures for a C-130. Basic rules of thumb for the C-130H variant are that the plane flies 300 knots per hour and consumes 5,000 pounds of fuel per hour. Its weight when empty is 85,000 pounds; its maximum takeoff weight is 155,000 pounds; its maximum fuel loading is 60,000 pounds. So except in the case of very light loads, adding more cargo requires a reduction in fuel—and hence range—in a linear, direct way. For example, if the payload were to weigh 25,000 pounds, 45,000 pounds of fuel could be carried, allowing about nine hours of flight or about 2,700 nautical miles of range. If the payload were reduced by 10,000 pounds, to 15,000 pounds total, then 10,000 more pounds of fuel could be loaded aboard—translating into two more hours and 600 more nautical miles of flight.[10]

Relevant data for a few other aircraft are as follows. Maximum takeoff weights (and maximum cargo capacities) are roughly 170 tons (and 30 tons) for the C-141, 290 tons (and 65 tons) for the C-17, 295 tons (and 60 tons) for the KC-10, and 420 tons for the C-5 or a 747B (with respective maximum cargo loadings of 89 and 100 tons). Fuel burn rates are 5.3 tons/hour for the C-141, 7 tons/hour for the C-17, 8.6 tons/hour for the KC-10, and 10.3 tons/hour for the C-5.[11] Maximum ranges at full payload are 2,500 miles for the C-141, 2,750 miles for the C-17, 4,400 miles for the KC-10, 7,300 miles for the C-5, and 8,300 miles for a 747-400.[12] This information allows one to do range-payload tradeoff calculations for various planes. For example, reducing the C-17's payload from sixty-five tons to thirty-five tons allows the addition of thirty more tons of fuel (all within the same maximum takeoff weight ceiling), which allows about four more hours of flight or 2,000 more miles of range.

Roll-on/roll-off ships can be loaded in three to four days and unloaded in two to three. Their speeds vary from 28 to 30 miles per hour for the SL-7 and large medium-speed roll-on/roll-off (LMSR) ships to 18 miles per hour for many ships in reserve. Most roll-on/roll-off ships have average payloads of about 15,000 to 20,000 tons (though their capacity is often constrained more by square footage; SL-7 ships have about 150,000 square feet of space, LMSRs about 250,000). Ships requiring cranes to move equipment on and off (including container ships) can take four to ten days for loading and a comparable amount of time for offloading. Their payloads vary greatly.[13]

Putting all these numbers together, it typically takes two to six large ships to move an entire division, depending on the division's size and weight. Alternatively, if it is to be flown (virtually never the approach taken for a whole division), the total number of flights required is typically 1,000 to 3,000 (depending not just on the type of division but the type of aircraft utilized, naturally).[14]

The United States has roughly 360 large airplanes for carrying troops and equipment (and another 200 quickly available via the civil reserve air fleet program). It also has about twenty large "roll-on roll-off" ships, each capable of carrying 15,000 to 20,000 tons of equipment (equipment and initial supplies for a heavy division weigh about 100,000 tons) as well as various other sealift ships. Altogether, its transportation assets create a theoretical capacity for a sustained average movement of about 30,000 tons of military equipment a day to a typical overseas destination.[15] To put it differently, assuming optimal rates, the United States could deploy about five divisions of ground forces, ten wings of Air Force combat aircraft, and associated initial support and supplies—perhaps totaling one million tons overall—in just over a month's time (though as a practical matter it would usually take several weeks to reach the level of maximum delivery, meaning closer to two months could be needed).

Constraints and Practical Obstacles

The preceding provides the theory, anyway. But in practice, deployments are often slower. Indeed, they can be much slower depending on the region to which troops, equipment, and supplies are being sent.

Bottlenecks often develop at ports and airfields, particularly abroad. Many airfields have very limited space for loading and unloading aircraft, for example, and perhaps limited refueling capacity as well, severely limiting throughput. Even at reasonably large, modern airfields it is generally difficult to deploy much more than 1,000 tons of equipment and supplies a day.[16] Absent at least three to four ports and airfields in the theater of destination, therefore, maximum deployment rates would not be attainable. In fact, actual deployment rates are often less than half of what is mathematically feasible in the abstract. That means the United States could need a few weeks to deploy a division-sized force and a few months to deploy a large force to most parts of the world.[17]

Indeed, transportation throughputs can be much less to certain types of locations. One careful analysis of possible military responses to the 1994 Rwanda genocide argued that it would have been difficult to deploy more than about 800 tons of military equipment and supplies a day to Rwanda using the two major airfields at Kigali and Entebbe, Uganda. The author, Professor Alan Kuperman, also noted possible constraints from issues such as ensuring adequate aerial refueling in the vicinity of Greece for transport aircraft. Factoring in the delays in getting started, and then in moving U.S. forces out of Kigali and about the country, he argued that a minimum of three weeks would therefore have been needed to deploy a reinforced brigade with 10,000 tons of equipment and 6,000 troops—longer than many assumed. Even if his mathematics can be challenged somewhat—for example, on the possible use of additional airfields in the region—the basic point about logistics and transport is correct. It would typically take a couple weeks even to deploy a few thousand troops to distant regions. The only possible exceptions to this general rule would be for cases where major U.S. or friendly bases were located nearby—or for cases where troops could be airdropped (in which case their tactical mobility would typically be very limited and their need for being resupplied could pose challenges).[18]

The Congressional Budget Office (CBO) is somewhat more optimistic than Kuperman in its analysis of a different case. CBO estimated that it could take about twenty-three days to deploy an existing heavy brigade to East Africa if only one airfield were available to receive incoming C-17 flights (the duration of the deployment is expected to decline somewhat, to eighteen to twenty days, if a new type of brigade based on the so-called future combat system and weighing about 25,000 tons is developed). This calculation assumes the maximum number of aircraft on the ground (or MOG) for an illustrative airport would be three, that a C-17 would require just over three hours to unload and get back in the air, that the airfield could be used on average twenty hours a day—and thus that, on average, sixteen flights per day could be handled.[19]

A final point: in addition to needing lots of assets, logistics requires lots of people. That is one reason why the Army has about twice as many soldiers dedicated to general support missions as to its main combat formations. In addition, it explains why tens of thousands of contractors are often also needed for large operations.[20] Indeed, even though the Army has about twice as many personnel in support as in combat units, its own estimates suggest it should have even more support—about 2.5 soldiers for every one in a combat unit.[21] The fact it does not is one reason so many contractors have been hired for the ongoing missions in Iraq and Afghanistan.

Basing

Understanding the U.S. military network can be daunting, given the dozens of countries and hundreds of facilities involved. But the number of major bases is actually rather modest. And for present purposes, that allows a complex situation to be greatly simplified. Many small American bases abroad are designed to create a symbolic presence, or facilitate a training mission with a host country, or provide a concrete manifestation of the strength of an alliance. In other words, while serving real and important purposes, they are not crucial for creating a global base network to facilitate large-scale military deployments and major regional operations. Only the latter activities are of primary concern here.

This section first summarizes the American base network abroad with an emphasis on crucial bases possessing broad regional and global importance. It then considers several policy issues of relevance to the network's future characteristics.[22]

Overview of America's Global Base Network

Regionally, American forces abroad are concentrated in three main zones—Europe, East Asia and the Western Pacific, and the broader Middle East including the Persian Gulf as well as Afghanistan. As of mid-2007, the United States had some 90,000 uniformed personnel assigned to bases in Europe (though about 10,000 of the total were deployed to the Central Command theater at that time, leaving some 80,000 actually in Europe). That was down from 120,000 in mid-2001 (with most of the largest reductions to date having been in Germany as well as the Balkans).[23]

The United States Air Force has a large presence in Europe, with just over 30,000 uniformed personnel employed there. The U.S. Air Force in

Europe operates seven main bases along with seventy smaller locations. The main operating bases are the Royal Air Force Lakenheath and Mildenhall Air Bases in England; Ramstein and Spangdahlem Air Bases in Germany, Aviano Air Base in Italy, Lajes Air Base in the Azores, and Incirlik Air Base in Turkey.

Incirlik in south central Turkey, after having hosted U.S. combat aircraft and more than 3,000 Americans during no-fly-zone operations over northern Iraq (Operation Northern Watch), has been downsized since the invasion of Iraq to a total of some 1,500 Americans that primarily support logistics and resupply flights. It is still a busy base given the amount of U.S. traffic going eastward from Europe, but on a substantially smaller scale than before.[24] In Germany, Ramstein Air Base is also a logistics hub, with an airlift wing as its core permanent unit.[25] Spangdahlem Air Base hosts F-16 and A-10 combat aircraft.[26] In Italy, Aviano Air Base hosts several dozen F-16 combat aircraft (it was critical in the air war against Serbia in 1999).[27] Lajes Field in the Azores Islands of Portugal is an important transit hub for many military aircraft crossing the Atlantic.[28] Finally, in the United Kingdom, Lakenheath is home to F-15 combat aircraft and Mildenhall to refueling aircraft.[29] Several hundred tactical nuclear weapons are still believed to be in Europe as well, distributed across bases in the U.K., the Netherlands, Belgium, Germany, Italy, Greece, and Turkey.[30]

U.S. naval facilities in Europe are found primarily in Spain and Italy (there are smaller capabilities in Germany, largely to help with port operations for loading and unloading forces, as well as Greece and Iceland). In Spain, the key facility is U.S. Naval Station Rota, on the Atlantic Ocean side of the Straits of Gibraltar. It is a support base for resupply, repair, and related activities for the Sixth Fleet. The Sixth Fleet headquarters is in Naples, Italy and another support base is found on the Italian island of Sicily.[31] A similar type of logistics hub is in Greece at Souda Bay.[32] There are all told about 8,000 U.S. sailors in Europe at these facilities.

The U.S. Army presence in Europe involves dozens of bases, many of which are being downsized or closed. Its drawdown in Europe is about halfway done. The number of soldiers was about 64,000 early this decade; it is down to about 45,000 now, and reportedly headed to 28,000 under current plans, though that figure seems likely to increase somewhat at least for a time, given the planned expansion in the Army.[33]

The Army is planning to deploy about a brigade's worth of troops to Eastern Europe at any given time, spread between Romania and Bulgaria on temporary deployments. A total of seven bases, all relatively near the

Black Sea, would be available for such purposes in the two countries, under the framework of ten-year agreements signed in 2005 with Romania and 2006 with Bulgaria. Smaller Army deployments have begun in 2007 (indeed, several U.S. Air Force combat aircraft deployed temporarily to Romania in 2006).[34]

In the Asia–Pacific region, the United States had about 74,000 uniformed personnel in mid-2007, down from a total of roughly 92,000 in mid-2001, with reductions in Korea being the most important change to date.[35] This is the region where American forces are the most evenly balanced by service, relative to other places with a large U.S. overseas presence—that 74,000 figure includes about 21,000 soldiers, 15,000 sailors, 16,000 Marines, and 22,000 Air Force personnel.[36]

In the Asia–Pacific theater, the dominant locations of American forces are in Japan and South Korea. Each has a formal countrywide U.S. military headquarters. The Pentagon's regional posture also includes important access to sites or collaborative training in Australia, Singapore, the Philippines, and elsewhere.[37] As for the main sites, U.S. Forces/Korea is focused virtually exclusively on the defense of the Republic of Korea; U.S. Forces/Japan is, by contrast, a regional and global hub.

In Japan, key Air Force bases reside in the north on the main island of Honshu (Misawa Air Base), as well as Yokota near Tokyo (home to the so-called Fifth Air Force) and Kadena Air Base on the island of Okinawa. The U.S. combat aircraft in Japan include the most modern variants of the nation's F-15 and F-16 fighters.[38] The Navy stations an aircraft carrier and air wing in the general vicinity of Tokyo (with ships at Yokosuka Naval Base and aircraft at Naval Air Facility Atsugi). That carrier is the only U.S. Navy aircraft carrier homeported abroad. More than 15,000 U.S. Marines are usually located on Okinawa, with key facilities including Camp Courtney, Camp Schwab, Camp Foster, Camp Butler, the Northern Training Area, and the Marine Corps Air Station Futenma, as well as Iwakuni on the main island of Honshu.[39] Major changes are planned for those Marines, such as moving about half of them (including those in the headquarters of the 3rd Marine Expeditionary Force [III MEF]) to Guam and relocating the Futenma Marine Corps airfield to a different and less populated part of Okinawa. (The Guam relocation plan is in its very fledgling stages, and is not due to be completed until 2014.)[40]

U.S. capabilities in Korea are focused primarily on the Air Force and Army, organized respectively into what the United States for largely historical reasons calls the 7th Air Force and 8th Army. The former has two main combat bases, Osan Air Base (only fifty miles from the DMZ, and home to the 51st Fighter Wing) and Kunsan Air Base (further south on Korea's west coast, and home to the 8th Fighter Wing). Kunsan features primarily F-16 aircraft (being upgraded from the so-called Block 30 to the more modern Block 40 configuration).[41] Osan has both F-16 and A-10 aircraft. Together, they host about 10,000 U.S. uniformed personnel.[42]

The other almost 20,000 American troops in Korea are mostly Army, centered on the 2nd Infantry Division (which despite its name is actually fairly heavy in terms of vehicles and armament). But other key units include the 19th Sustainment Command and Logistic Support Element Far East (both of which would help with the flow of hundreds of thousands of additional U.S. troops to the peninsula in an all-out war), as well as special operations forces and the 35th Air Defense Artillery Brigade (which fields Patriot missile defense systems among other capabilities).

Beyond Japan and Korea, American capabilities in Alaska and Hawaii, while of course on U.S. territory, also occupy a hybrid status of sorts—constituting forces on U.S. territory that are also to some degree forward deployed. That is also true for the growing presence on Guam, which is soon to feature three attack submarines, up to forty-eight fighter aircraft, up to ten Global Hawk spy planes, special forces, tanker aircraft, Navy vessels known as Littoral Combat Ships, and those 8,000 Marines from Okinawa.[43] Many of these aircraft may also have hardened shelters built for them (and over time, hardened runways are a possibility, too, depending on how the theater evolves). It also has the capacity for massive reinforcement; up to 170 B-52 bombers at a time operated there during the Vietnam War.[44]

The United States is losing an air base in Ecuador that has been used by several aircraft such as AWACS to maintain surveillance over and near that country (including over ocean waters) as part of the drug war. The loss of this base is potentially significant, to be sure—but for its direct localized effects on narcotics interdiction, not for broader or larger military operations in the waters around Ecuador or beyond. As such, it is a secondary matter in the context of this analysis of America's global base network and how that allows power projection around the globe.[45]

Moving to the Middle East, U.S. military capabilities are of course found overwhelmingly in Iraq at present. There were about a dozen very large bases, and some forty-five major bases overall, as of the peak of operations in 2007. Counting forward-operating bases and combat outposts, the number of installations exceeded 100. The larger bases include Camp Victory at the Baghdad Airport, where the main U.S. military headquarters as well as two American divisions were located (as of 2007), Camp

Anaconda/Balad Air Base north of Baghdad (home to the 332nd Air Expeditionary Wing, the only Air Force wing in Iraq), and Camp Speicher near Tikrit.[46]

Kuwait hosts the second largest U.S. capability in the region, with some 20,000 troops. Roughly sixteen bases are currently used to support this presence.[47] Bahrain is notable for providing a home to the U.S. Fifth Fleet, with more than 1,000 American sailors located there. Qatar is a major logistical hub as well as the regional headquarters for Central Command (the main headquarters for which are in Tampa, Florida) and command center for the main U.S. regional command airbase (Al Udeid); several hundred Americans are stationed in that country. Egypt, a major non-NATO ally, allows invaluable air and naval transit for U.S. forces through the Suez Canal. It is also home to the biennial BRIGHT STAR multinational training event for CENTCOM. Some 400 U.S. personnel are in Egypt, mostly airmen and airwomen. Also key in the war on terror is a combined joint task force operating out of Djibouti for the Horn of Africa. Almost 1,500 U.S. troops are stationed there, apportioned roughly equally among the military services.[48] About 1,000 U.S. personnel are also found on Diego Garcia, a British-owned territory in the Indian Ocean, a major logistical hub.

The United States has about 2,400 uniformed personnel in sub-Saharan Africa (as of mid-2007), up substantially from the 300 there in mid-2001.[49] There were thirteen cooperative security locations in Africa under EUCOM jurisdiction prior to the creation of AFRICOM. With that command's creation, the cooperative security locations switch to its jurisdiction—though the fact that AFRICOM is based in Europe, with an uncertain political status and acceptability in much of Africa itself, provides more questions than answers about the future of U.S. capabilities on that continent.[50]

Normally, the Army has two brigades' worth of prestationed or prepositioned combat equipment in Kuwait as well. However, ongoing operations in Iraq and Afghanistan have largely made use of these (as well as two brigade sets typically on ships, one at Guam, the other on the Indian Ocean island of Diego Garcia). Of the Army's normal allotment of five prepositioned brigade sets of equipment, only the set in South Korea is roughly complete at present.[51] For these reasons, rather than being able to make several brigades operational within as little as five to ten days (as troops were airlifted to join up with their equipment), the United States would have to load up equipment from the United States and then ship it across the oceans, necessitating a month or more even if equipment was promptly available for loading. This situation would have posed a substantial risk in the decades when North Korea's military was stronger than South Korea's; it is probably less worrisome now, but still a strategic constraint on prompt American responsiveness to various places.

The Marine Corps has a policy of keeping a brigade's worth of prepositioned equipment afloat at Diego Garcia, a second in the Mediterranean, and a third at Guam. The Air Force keeps ammunition ready to move quickly on ships at Diego Garcia; the Army also keeps support equipment quickly deployable on Guam.[52]

Purposes and Functions of Major Bases

How to understand the various roles played by these bases? Even among the major facilities, there exists a wide range of main missions and objectives. They include:

- Combat bases for tactical aircraft (as in Japan and Korea)
- Ground combat bases (Iraq and Afghanistan and central parts of Korea)
- Ports for ships (as in Japan and Guam, and to a lesser extent Bahrain, Diego Garcia, and Italy)
- Major logistics hubs for equipment repair, storage and transshipment of supplies and equipment, refueling of aircraft (as in Germany, Kuwait, Turkey, Italy, Spain, Diego Garcia, southern Korea and again Guam and Japan)
- Ground force bases focused on maintaining a forward presence for general deterrence and for alliance exercises (Germany and Okinawa)

Which of these functions are most significant? It is important to ask in case choices might have to be made about which to keep (should a host nation, for example, ask for a smaller U.S. footprint due to domestic political realities). It is also important to develop backup options for those bases that are truly crucial. Moreover, which bases are most likely to survive in the face of a possible enemy attack? This is a growing worry in an era of increasingly precise ballistic and cruise missile inventories for many countries.

Working through the same list, combat air and ground bases are clearly quite important if one anticipates the real possibility of fighting a major war (or deterring it) in a given region. That makes combat air bases in

Japan and Korea quite important but suggests that remaining tactical air bases in places such as Germany and Britain are less important, given the stable character of contemporary Western Europe. For tactical combat aircraft, it is optimal to have bases within about 500 miles of where they are likely to conduct wartime operations; facilities in Japan and Korea are within that distance of North Korea and the Taiwan Strait, whereas those in Western Europe probably are not close to any likely future combat theater.

How many air bases are needed in a given location? It is generally difficult to operate more than a wing of aircraft (about seventy-two planes) from a given base, based on historical averages and air traffic practicalities. So several major air bases, say at least ten, would ideally be available for a major operation given the experiences of the Iraq wars.

This might seem a fairly straightforward number to have available, as the world inventory of runways of at least 6,000 feet with sufficient strength to handle at least a fighter jet exceeds 2,000 (roughly 300 in Asia, 150 in the Persian Gulf and Middle East region, 400 in Western Europe, 600 in North America, and about 550 in the rest of the world combined). But they are, of course, unevenly distributed. Many countries only have one or two (and those with hardened shelters for aircraft that might be attacked are far fewer).[53] Moreover, the United States only has about fifteen major air bases abroad in a total of ten countries now. So there is a premium on retaining access to these facilities.[54] Aircraft carriers, and improvised use of commercial airfields or other countries' military airfields, can clearly help in crises if available. But they are generally less optimal for sustained combat operations.

One reason to have a certain number of dedicated military airfields concerns airbase survivability. Aircraft, runways, and the people working on airplanes or flying them are increasingly vulnerable to precision weapons and submunitions (not to mention attacks by special forces, which have damaged or destroyed more than 2,000 aircraft since 1942, according to Alan Vick of RAND).[55] Israel proved this even in 1967 when it destroyed 300 Egyptian aircraft within five hours. The Israelis had time for a second strike in 1967 because in their first wave of attacks they had also used runway-penetrating weapons that prevented Egypt from moving its surviving aircraft.

Runway repair equipment and multiple runway surfaces can help address the vulnerability problem by making it easier to restore flight operations quickly. Hardened facilities and shelters for tactical aircraft can help too, as Egypt proved in 1973. Of course, to have hardened shelters and repair equipment available requires planning and a political environment in which such preparations are possible. Accordingly, while some 1,400 hardened shelters lie in Asia, almost half (some 640 as of 2002) are in the Republic of Korea and are not necessarily available for conflicts beyond the peninsula. Also, about 100 hardened shelters are in Japan, at least 200 in Taiwan, just over 200 in India, and just under 200 in Pakistan—but which if any would be available to U.S. forces in a crisis or conflict is hard to say, and most would be well out of range of any given conflict.[56] Moreover, these shelters are expensive, to the tune of about $4 million a piece typically (with costs for construction of an entire modern air base likely to reach $1.5 billion or more, and the time period for construction likely to reach into many months at a minimum). Although modern precision weapons make many shelters vulnerable to direct hits in ways they were not years ago, shelters are still generally a great aid to aircraft survivability.[57] And repair equipment can keep runways and taxiways functioning. Such survivability assets are generally much easier to build and operate at dedicated military bases than elsewhere.

What about the Navy? Of course, American warships do not require coaling stations to operate globally any more. But they can maintain deployments more efficiently if homeported abroad. Such stationing reduces time wasted in transit; it also allows a given vessel to be essentially "on call" at all times, even when in port. This means, for example, that there is a huge benefit to the use of ports in Japan in particular (that is the main place where American warships are routinely home-based). By some estimates, to achieve the same continual level of presence and deterrence in a place like the Persian Gulf or Western Pacific, it could require four to six times as many ships operating out of American ports as operating from a well-positioned forward homeport.[58]

Logistics hubs are quite important, especially in regions where huge numbers of personnel and supplies might need to be surged in a major crisis. As noted, ports and airfields in combat theaters often get clogged up during big deployments; they can also be vulnerable to attack. And as with any base, access to it may be denied by a host government in a given crisis or war. All these factors place a high premium on having a number of bases available; they also place a premium on being able to set up infrastructure such as underground fuel storage facilities in advance of any crisis. Having logistics hubs where large ships and planes can be unloaded and supplies transferred to smaller, more manageable vessels and aircraft is critical for reducing bottlenecks and increasing safety in major deployment operations.

What about the last bullet point from the previous list, the symbolic value of bases? While not trivial, it is not generally the most important reason to have bases. There usually are alternatives to stationing large ground force contingents overseas if their main purpose is to show the flag and conduct exercises with allies. Forward presence is important, but it is also largely an intangible, and as such modest reductions in a given capacity can often be tolerated if need be.

QUESTION 11: For global use, which is a better place for the United States to station Army forces, Texas or Germany?

ANSWER: The key point to recognize in answering this question is, as shown in Table 3.1 and 3.2, that heavy forces are quite massive. They almost always require movement by sea; only small elements of them can realistically be delivered intercontinentally by air. The practical importance of this fact is that one must think through the steps of loading and sailing ships, not just look at a map and think of distances as the bird flies, when evaluating transportation options.

Seen in this light, having U.S. ground forces in Western Europe in the modern era may be of little benefit, even though those forces are closer to likely trouble spots in the Middle East than if based in Texas or Georgia or North Carolina or California. This is because the most efficient deployment path from Germany to the Persian Gulf is via rail to ports in the Baltic Sea, and then over the Atlantic Ocean and into the Mediterranean. Despite the fact that Germany has an excellent infrastructure and tends to facilitate the movement of American forces through its territory (even in the runup to the 2003 invasion of Iraq, when the German government opposed the war), that trajectory is little better than a deployment from the United States (and it is worse than starting from the central or western United States if the goal is to reach East Asia). Perhaps three days of sailing time would be averted en route to the Gulf, at most; other timelines, for loading forces on trains and moving them to ports, as well as loading and unloading ships, would be essentially unchanged. In fact, deployment from Europe could actually be *slower* unless sealift ships were already on hand in Europe when a crisis began and the clock started ticking.

It is important to understand, in considering such options, that the savings associated with basing forces in the United States rather than abroad are generally modest. Costs to purchase equipment, train, pay salaries, and the like are comparable whether forces are stationed at home or abroad; some base costs may also be partially covered by allied nations (Japan is especially generous in this regard). As such, bringing home 50,000 soldiers from Europe and South Korea (the total number deployed in those two countries as of 2004, before recent reductions began, was 80,000) would save about $1 billion a year in annual operating costs once implemented (about half in base operations, half in personnel costs for purposes such as moving people about and running schools abroad). The bases themselves that would be rendered unnecessary by such a change are worth billions. But they would generally revert back to their host nations rather than be available for sale by the United States. In fact, if replacement bases had to be rebuilt in the United States, costs for stationing more forces in the United States would be greater than those for keeping them abroad for a time. That fact helps explain why the U.S. military is now slowing its reductions in Europe, since its recent decision to increase the size of the Army and Marine Corps makes available basing back home somewhat scarce and would necessitate the construction of more facilities for returning units.[59]

QUESTION 12: Which U.S. bases in Japan are most important?

ANSWER: It is useful to break this question down into two main parts: the large (and controversial) U.S. military presence on Okinawa, and the presence on the main island of Honshu (focused on the homeporting of a U.S. aircraft carrier battle group in Yokosuka, though there are other important U.S. military capabilities on the main islands, too, including more aircraft in the north of Japan and Navy ships near Nagasaki).

Tactical air bases are extremely important in Okinawa, and in Japan and Korea more generally, to reach potential combat zones in North Korea and the Taiwan Strait. There are only two main U.S. tactical air bases on Okinawa—Kadena for the Air Force and Futenma for the Marine Corps—underscoring their importance. (Some of that extreme importance may have been mitigated by recent U.S.-Japan agreements to make available Japan's own military and civilian airfields in the event of a crisis, but these facilities are less well prepared for supporting combat operations.) The other alternative, in the event of war, would be to build new air bases in the course of conflict. While bulldozing and paving can be done fairly fast, within weeks perhaps under some circumstances, construction of the vast underground fuel and ammunition storage facilities needed for such operations as well as hardened shelters for the aircraft themselves realistically requires months. Not having immediate access to air bases thus makes forward defense less credible and weakens deterrence.

The U.S. Marine Corps presence on Okinawa, by contrast, is more a matter of convenience than of military necessity—as evidenced by the fact that the United States is now willing to bring half of those Marines to Guam to reduce the associated political burden on Okinawan and Japanese politicians. As noted, the Marines do not fight on Okinawa; they would have to load up on ships to deploy in substantial numbers with their equipment (and there are only enough amphibious ships in Japan to deploy about 2,000 of them with their combat equipment at a time). Aerial transport can move modest numbers of Marines, but equipment and supplies are so heavy that the throughput capacity would be limited, as Tables 3.2 through 3.5 underscore. Okinawa for the Marines is mostly a staging base, not a combat base. The Marine Corps presence on Okinawa may have important symbolic virtue, but its military necessity probably does not rival that of the Air Force Kadena base there.

Moving to the aircraft carrier issue, consider the hypothetical of the homeport being lost. In that situation, the most straightforward (though expensive) response would be to build a larger Navy to compensate and base the ships in the United States. Then, the same number of days per year of needed forward presence could be maintained, but only by virtue of having more ships share in the task. Typically, for ships homeported in the United States, the Navy will be able to maintain a ship forward deployed and on station about 20 to 25 percent of the time. It will use another 25 percent for training, perhaps 10 to 15 percent for moving the ship to and from the overseas theater in question, another 25 percent for rotating crews and allowing recovery time on shore, and an average of another 10 to 15 percent for major ship overhaul every few years. So a forward presence of one carrier in the Western Pacific could be maintained either by one carrier homeported in Japan (and considered to be on station at all times) or about five carriers sharing the job and based back in the United States. Given the costs of aircraft carriers (see Table 1.5), the difference could be as much as $25 billion a year, if it really was necessary to construct a larger Navy to compensate for the loss of Yokosuka.

Notes

1. Department of Defense, *Conduct of the Persian Gulf War: Final Report to Congress* (Washington, D.C.: Department of Defense, April 1992), pp. F-1, F-2, F-26.

2. Thomas A. Keaney and Eliot A. Cohen, *Gulf War Air Power Survey Summary Report* (Washington, D.C.: Government Printing Office, 1993), p. 210.

3. Rachel Schmidt, *Moving U.S. Forces: Options for Strategic Mobility* (Washington, D.C.: Congressional Budget Office, 1997), p. 80.

4. Frances M. Lussier, *Replacing and Repairing Equipment Used in Iraq and Afghanistan: The Army's Reset Program* (Washington, D.C.: Congressional Budget Office, September 2007), pp. 1–5.

5. Rachel Schmidt, *Moving U.S. Forces: Options for Strategic Mobility* (Washington, D.C.: Congressional Budget Office, February 1997), p. 48.

6. David A. Ochmanek, Edward R. Harshberger, David E. Thaler, and Glenn A. Kent, *To Find, and Not to Yield* (Santa Monica, Calif.; RAND, 1998), p. 27; Christopher Bowie, Fred Frostic, Kevin Lewis, John Lund, David Ochmanek, and Philip Propper, *The New Calculus: Analyzing Airpower's Changing Role in Joint Theater Campaigns* (Santa Monica, Calif.: RAND, 1993), pp. 30–33; Thomas A. Keaney and Eliot A. Cohen, *Gulf War Air Power Survey Summary Report* (Washington, D.C.: Government Printing Office, 1993), pp. 210–13; and Christopher J. Bowie, *The Anti-Access Threat and Theater Air Bases* (Washington, D.C.: Center for Strategic and Budgetary Assessments, 2002), pp. 15–16.

7. Personal communication with Lt. Col. Chris Patterson, U.S. Air Force, Scott Air Force Base, U.S. Transportation Command, Belleville, Illinois, February 21, 2008.

8. Button, Gordon, Riposo, Blickstein, and. Wilson, *Warfighting and Logistic Support of Joint Forces from the Joint Sea Base*, pp. 99–101; and Colonel Timothy M. Laur and Steven L. Llanso, *Encyclopedia of Modern U.S. Military Weapons* (New York: Berkley Books, 1995), pp. 116–20, 150.

9. U.S. Air Force, "Air Force Fact Sheet: C-130 Hercules," May 2006, available at www.af.mil/factsheets/factsheet.asp?id=92 [accessed December 14, 2007].

10. Conversation with Lt. Col. William Knight, U.S. Air Force (and Congressional Research Service fellow, 2007–2008), January 2, 2008.

11. Typical cargo loadings are about two-thirds of the theoretical maximum; tons here are short tons, or 2,000 pounds. See U.S. Air Force, *Air Mobility Planning Factors*, Air Force Pamphlet No. 10-1403 (June 1997), available at www.fas.org/man/dod-101/usaf/docs/afpam10-1403.htm [accessed February 19, 2008].

12. See Federation of American Scientists, "C-141B Starlifter," Washington, D.C., 1999, available at www.fas.org/man/dod-101/sys/ac/c-141.htm [accessed February 19, 2008]; U.S. Air Force, "Fact Sheet: C-17 Globemaster III," Washington, D.C., May 2006, available at www.af.mil/factsheets/factsheet_print.asp?fsID=86&page=1 [accessed February 19, 2008]; U.S. Air Force, "Fact Sheet: KC-10 Extender," Washington, D.C., September 2006,

available at www.af.mil/factsheets/factsheet_print.asp?fsID=109&page=1 [accessed February 19, 2008]; U.S. Air Force, "Fact Sheet: C-5 Galaxy," Washington, D.C., August 2007, available at www.af.mil/factsheets/factsheet_print.asp?fsID=84&page=1 [accessed February 19, 2008]; and AerospaceWeb.Org, "Boeing 747 Long-Range Jetliner," February 2008, available at www.aerospaceweb.org/aircraft/jetliner/b747/ [accessed February 19, 2008].

13. Arthur, *Options for Strategic Military Transportation Systems*, pp. 5, 8, 14.

14. Schmidt, *Moving U.S. Forces*, p. 80.

15. This estimated capacity for sustained delivery from airlift and sealift together is not to be confused with a metric commonly used for airlift in particular, million ton miles per day (MTM/D). The United States presently has nearly sixty MTM/D of airlift capacity—defined as the sum of all airlifters' payload, times their average speed, times their average number of sustainable hours of flight per day, all divided by two to account for the fact that the planes must fly back empty (more or less) to load up again for another trip. See David Arthur, *Options for Strategic Military Transportation Systems* (Washington, D.C.: Congressional Budget Office, September 2005), pp. 8–9.

16. Schmidt, *Moving U.S. Forces*, pp. 48, 54, 80–81; and Department of Defense, *Conduct of the Persian Gulf War: Final Report to Congress* (Washington, D.C.: Department of Defense, April 1992), p. F-26.

17. Arthur, *Options for Strategic Military Transportation Systems*, pp. x, xii, 3, and 5.

18. Alan J. Kuperman, *The Limits of Humanitarian Intervention* (Washington, D.C.: Brookings, 2001), pp. 54–77 (especially pp. 60–61).

19. Frances M. Lussier, *The Army's Future Combat Systems Program and Alternatives* (Washington, D.C.: Congressional Budget Office, 2006), pp. 31, 36, 53, 71–72.

20. Matthew Goldberg, *Logistics Support for Deployed Military Forces* (Washington, D.C.: Congressional Budget Office, October 2005), pp. 2–5, 17. The Active Army in 2005 had 151,000 soldiers in main combat units, 79,000 in combat support (capabilities such as air and missile defense), and 92,000 in what is termed combat service support (such as rear-area logistics). For the National Guard, the respective numbers were 169,000, 67,000, and 89,000. For the Army Reserve, they were 14,000, 40,000, and 84,000. Overall totals for the entire Army were thus 334,000 in combat units, 187,000 in combat support, and 265,000 in combat service support. And for contractors, in Operation Desert Shield and Desert Storm more than 60,000 contractors were employed in the theater (for maintenance, cargo trucking, engineering, fuel supply, and other purposes). In Operation Iraqi Freedom, nearly 200,000 have been employed, including non-Americans.

21. Frances M. Lussier, *Structuring the Active and Reserve Army for the 21st Century* (Washington, D.C.: Congressional Budget Office, 1997), pp. 10–11.

22. Some of the ideas in this section first appeared in Michael E. O'Hanlon, *Unfinished Business: U.S. Overseas Military Presence in the 21st Century* (Washington, D.C.: Center for a New American Security, 2008).

23. Department of Defense, "Active Duty Military Personnel Strengths by Regional Area and by Country," June 30, 2007, available at siadapp.dmdc.osd.mil/personnel/MILITARY/history/hst0706.pdf [accessed October 30, 2007]; and Department of Defense, "Active Duty Military Personnel Strengths by Regional Area and by Country," June 30, 2001, available at siadapp.dmdc.osd.mil/personnel/M05/hst0601.pdf [accessed June 30, 2007].

24. U.S. Air Force, "Incirlik Air Base," available at www.incirlik.af.mil/units [accessed June 26, 2007]; and Alparslan Akkus, "Incirlik Shop Owners Moved to U.S. Base in Iraq," *Turkish Daily News*, April 28, 2007, available at www.turkishdailynews.com/tr/article.php?enewsid=71815.

25. See U.S. Air Force, "Ramstein Air Base," available at www.ramstein.af.mil/units [accessed June 27, 2007].

26. See U.S. Air Force, "Spangdahlem Air Base," available at www.spangdahlem.af.mil/units [accessed June 27, 2007].

27. U.S. Air Force, "Aviano Air Base," available at www.aviano.af.mil/units [accessed June 27, 2007].

28. U.S. Air Force, "Lajes Field," available at www.lajes.af.mil/units [accessed June 27, 2007].

29. U.S. Air Force, "Royal Air Force Mildenhall," available at www.mildenhall.af.mil/units [accessed June 27, 2007]; and U.S. Air Force, "Royal Air Force Lakenheath," available at www.lakenheath.af.mil [accessed June 27, 2007].

30. Natural Resources Defense Council, "U.S. Nuclear Weapons in Europe: A Review of Post–Cold War Policy, Force Levels, and War Planning," New York, NY, February 2005, Appendix A, available at nrdc.org/nuclear/euro/euro_app.pdf [accessed November 2, 2007]. NRDC estimates that the United States had a total of 480 nuclear weapons in Europe in early 2005.

31. Sandra Jontz, "Navy Base at La Maddalena May Close Months Earlier," *Stars and Stripes*, January 21, 2007, available at stripes.com/article.asp?section=104&article=41844&archive=true [accessed July 10, 2007].

32. See U.S. Navy, "Welcome to Rota," available at www.rota.navy.mil/navsta/welcome/virtual_tour/text_only.html [accessed June 27, 2007]; Charlie Coon, "Transformation: Navy Shifts Its Priority Away from the North Atlantic," *Stars and Stripes*, June 28, 2007, available at www.estripes.com/article.asp?section=1048article=543798archive=true [accessed June 28, 2007]; and Ben Murray, "Navy Readies Closing of U.K. Command," *Stars and Stripes*, July 25, 2007, available at stripes.com/article.asp?section=104&article=52316&archive=true [accessed July 26, 2007].

33. Leo Shane III, "Transformation: Stateside Bases Prepare for Influx," *Stars and Stripes*, June 20, 2007, available at stripes.com/article.asp?section=104&article=54363&archive=true [accessed July 14, 2007].

34. Charlie Coon, "Soldiers in Italy, Germany Bound for Romania," *Stars and Stripes*, July 10, 2007, available at www.estripes.com/article.asp?section=104&article=54789&archive=true; and Bryan Mitchell, "USAFE Strengthens Partnership with Romanians," *Stars and Stripes*, July 11, 2007, available at www.estripes.com/article.asp?section=148&article=53654&archive=true [accessed July 27, 2007].

35. Department of Defense, "Active Duty Military Personnel Strengths by Regional Area and by Country," June 30, 2001, available at siadapp.dmdc.osd.mil/personnel/M05/hst0601.pdf [accessed June 30, 2007].

36. Department of Defense, "Active Duty Military Personnel Strengths by Regional Area and by Country," June 30, 2007, available at siadapp.dmdc.osd.mil/personnel/MILITARY/history/hst0706.pdf [accessed October 30, 2007].

37. Statement of Admiral William J. Fallon, United States Navy, and (then) Commander, United States Pacific Command, before the House Armed Services Committee, March 7, 2007.

38. John A. Tirpak, "Comeback in the Pacific," *Air Force Magazine* (July 2007), p. 26.

39. See Department of Defense, "Welcome to U.S. Forces, Japan," and "USFJ Fact Sheet," available at www.usfj.mil [accessed May 15, 2007].

40. Secretary of State Condoleeza Rice, Secretary of Defense Robert Gates, Minister for Foreign Affairs Taro Aso, and Minister of Defense Fumio Kyuma, "Joint Statement of the Security Consultative Committee, Alliance Transformation: Advancing United States–Japan Security and Defense Cooperation," May 1, 2007; and U.S. Forces, Japan, "Welcome to U.S. Forces, Japan," [accessed May 15, 2007].

41. John A. Tirpak, "Comeback in the Pacific," *Air Force Magazine* (July 2007), p. 26.

42. See, for example, U.S. Air Force, "7th Air Force," available at www.7af.pacaf.af.mil/units [accessed May 18, 2007].

43. Christian Caryl, "U.S. Military Embraces Guam," *Newsweek International*, February 26, 2007, available at www.msnbc.msn.com/id/17202830/site/newsweek.

44. Tirpak, "Comeback in the Pacific," pp. 26–27.

45. Joshua Partlow, "Ecuador Giving U.S. Air Base the Boot," *The Washington Post*, September 4, 2008, p. A6; and Kintto Lucas, "Ecuador: Manta Air Base Tied to Colombian Raid on FARC Camp," IPS News, March 21, 2008, available at http://ipsnews.net/news.asp?idnews=41687 [accessed September 4, 2008].

46. See U.S. Air Force, "Air Force in Iraq, 332nd AEW," April 2007, available at www.balad.afnews.af.mil/library/factsheets/factsheet.asp?id=4032; Friends Committee on National Legislation, "Iraq," October 27, 2005, available at fcnl.org/iraq/bases_text.htm [accessed July 12, 2007].

47. Krepinevich and Work, *A New Global Defense Posture for the Second Transoceanic Era*, p. 149; and John Pike, "Kuwait Facilities," June 20, 2005, available at www.globalsecurity.org/military/facility/kuwait.htm [accessed August 10, 2007].

48. Statement of Admiral William J. Fallon, United States Navy, and Commander, United States Central Command, before the Senate Armed Services Committee, May 3, 2007.

49. Department of Defense, "Active Duty Military Personnel Strengths by Regional Area and by Country," June 30, 2001, available at siadapp.dmdc.osd.mil/personnel/M05/hst0601.pdf [accessed June 30, 2007].

50. Statement of General Bantz J. Craddock, U.S. Army and Commander, United States European Command, March 15, 2007, before the House Armed Services Committee, pp. 16–25.

51. Ann Scott Tyson, "Military Is Ill-Prepared for Other Conflicts," *The Washington Post*, March 19, 2007, p. A1; and Government Accountability Office, *Defense Logistics: Improved Oversight and Increased Coordination Needed to Ensure Viability of the Army's Prepositioning Strategy*, GAO-07-144 (February 2007), p. 17.

52. Schmidt, *Moving U.S. Forces*, pp. 36, 40; and Eric Labs, *The Future of the Navy's Amphibious and Maritime Prepositioning Forces* (Washington, D.C.: Congressional Budget Office, November 2004), p. 6.

53. Christopher J. Bowie, *The Anti-Access Threat and Theater Air Bases* (Washington, D.C.: Center for Strategic and Budgetary Assessments, 2002), pp. 17–18, 25, 46, 71–72.

54. Bowie, *The Anti-Access Threat and Theater Air Bases*, pp. 17–18, 23–24, 31.

55. Alan Vick, *Snakes in the Eagle's Nest: A History of Ground Attacks on Air Bases*, MR 553-AF (Santa Monica, Calif.: RAND Corporation, 1995).

56. Bowie, *The Anti-Access Threat and Theater Air Bases*, pp. ii–iii, 5–6.

57. Ibid., pp. 42–46, 54–55.

58. See Michael O'Hanlon, "Restructuring U.S. Forces and Bases in Japan," in Michael M. Mochizuki, ed., *Toward a True Alliance* (Washington, D.C.: Brookings, 1997), pp. 149–78.

59. Frances M. Lussier, *Options for Changing the Army's Overseas Basing* (Washington, D.C.: Congressional Budget Office, 2004), pp. xiv, 43, 52–54.

Key References and Suggestions for Further Reading

Arthur, David, *Options for Strategic Military Transportation Systems* (Washington, D.C.: Congressional Budget Office, September 2005).

Bowie, Christopher J., *The Anti-Access Threat and Theater Air Bases* (Washington, D.C.: Center for Strategic and Budgetary Assessments, 2002).

Button, Robert W., John Gordon IV, Jessie Riposo, Irv Blickstein, and Peter A. Wilson, *Warfighting and Logistic Support of Joint Forces from the Joint Sea Base* (Santa Monica, Calif.: RAND, 2007).

Labs, Eric, *The Future of the Navy's Amphibious and Maritime Prepositioning Forces* (Washington, D.C.: Congressional Budget Office, November 2004).

Lussier, Frances M., *Options for Changing the Army's Overseas Basing* (Washington, D.C.: Congressional Budget Office, 2004).

Schmidt, Rachel, *Moving U.S. Forces: Options for Strategic Mobility* (Washington, D.C.: Congressional Budget Office, 1997).

Vick, Alan, *Snakes in the Eagle's Nest: A History of Ground Attacks on Air Bases*, MR 553-AF (Santa Monica, Calif.: RAND Corporation, 1995).

CHAPTER IV

Technical Issues in Defense Analysis

How does an analyst or policy specialist wade into the complex world of actual military science—that is, the realm of physics and engineering on which so many practical decisions about military matters turn? Given the sophistication of the technologies involved, this would seem an impossible task for the generalist, or even for many scientists lacking specialized knowledge of certain aspects of military technical matters. But in another sense, it is a necessary task. Only by striving for answers to questions like can missile defenses work, can space weapons provide capabilities unavailable from systems based on Earth, and can future warfare be radically transformed by changes in underlying weapons systems and tactics may we reach decisions about proper defense resource allocation. Only in these ways can we, even more importantly, avoid major surprises in future wars (or, to put it differently, profit from any surprises before adversaries can do so). Only in these ways can we understand the potential, and the limits, of arms control.

This section of the book provides a primer on some of these key technical subjects to aid the general reader. In so doing, it suggests an analytical approach to addressing such subjects that may be of broader use even for matters not discussed here.

The utility of such a primer is limited in part by the technical proficiencies of the author, but even more fundamentally by the fact that technologies change with time, and that basic knowledge can only go so far in answering questions that often require detailed precise information about the very latest technological trends and opportunities. To pursue state-of-the-art scientific work, scientists are clearly needed, and policy generalists cannot be of great use. It took Szilard and Einstein to warn President Roosevelt that nuclear weapons were possible, for example, and it also took experts to figure out when capabilities like radar, aerial flight, space flight, and laser sensors as well as weapons were within reach. To anticipate future breakthroughs, and help decide which technologies are worth pursuing, the Department of Defense has numerous expert scientific advisory groups and consultants today—ranging from the notorious JASON group and the Defense Science Board, to the main weapons laboratories run by DoD (like Lincoln Labs at MIT) and the military services as well as the Department of Energy, to many individual scientists or groups of scientists working either for the defense industry or for universities.

That said, some basic understanding of scientific and technical issues in defense policy is essential for the policymaker. Scientists cannot be asked to make all decisions concerning technology, since many decisions involve other matters, too—the country's national security objectives, its resource constraints, its competing priorities, its arms control interests, and so on. Since many core matters in defense policy revolve heavily around physics and engineering, a basic familiarity with these fields is necessary. Even a very limited basic knowledge about key concepts and terminologies allows the generalist to follow conversations and studies led by more technically expert individuals. If generalists are at least able to follow technical discussions, they can often discern the key assumptions behind science-based arguments. In other words, basic scientific literacy among generalists helps create a vetting process that can often weed out sloppy, mistaken, or ideologically motivated arguments. It can also make generalists more able to appreciate the work of whistleblowers and dissidents from *within* the scientific community, and pay them heed when institutional and political forces may otherwise overwhelm them.[1]

Some basic matters of physics are both simple enough to be accessible to the generalist, and important and enduring enough that they can be expected to remain relevant for policymakers well into the future. When the immutable laws of physics can be invoked to help understand a situation, the resulting explanation is more likely to be durable. It is not always possible to find basic physical arguments or principles that help resolve a technical issue or debate, but it often is. Making some investment in understanding core physics concepts can then have benefits far down the road.

Some examples of how a sound understanding of basic military principles and technologies can inform policy debate may illustrate these points. Take the missile defense debate of the 1980s, shortly after Ronald Reagan's "Star Wars" speech of 1983 in which he announced his Strategic Defense Initiative (SDI). Whatever the broader strategic benefits of SDI may have ultimately been, many of the technical goals advanced by its partisans could be debunked—or shown to be very expensive and rather improbable—by basic physical reasoning. For example, putting lasers in space to shoot down warheads could be shown to require dozens of lasers (because of the Earth's rotation, meaning a given satellite would not stand still over a given point on the planet's surface). With each of those lasers requiring a mirror effectively equivalent to that of the Hubble telescope just to steer the beam, costs could be placed into the many tens of billions for just the initial deployment of the system (even assuming its technical feasibility). On a related subject of SDI, a good deal of analysis about possible countermeasures that an attacker could use to fool a defense suggested that any country sophisticated enough to build a substantial nuclear-tipped ballistic missile inventory could defeat most basic defenses. These arguments did not shut the door on all possible uses for missile defense, by any means, but they were sobering for those who wanted to believe that defense could trump offense in the nuclear realm.

In 1999, many predicted that NATO airpower could easily intimidate Serbian militias into stopping their deprivations against the Kosovar Albanian population. But others recognized that, if NATO planes stayed above 15,000 feet altitude to reduce their vulnerability, their ability to identify and target small Serb formations would be extremely limited—and the ability of their precision bombs to strike accurately through the cloud cover prevalent in the Balkans in early spring would be limited as well. Again, basic science, coupled with Clausewitzian cautions about fog and friction in war, provided policy-relevant insights.

Finally, the popular hypothesis of the 1990s and early years of this century that a revolution in military affairs (RMA) was underway led to a number of unsound predictions. Some knowledge of technology tended to make it easier to see why skepticism was warranted. Chief among the assertions of the RMA proponents was that most, if not all, forms of warfare would be radically changed, leading to a much different (and reduced) role for traditional ground power in combat. The technological basis for this prognostication was always weak, and it did not take a Ph.D. in physics to know why. Yet the RMA movement is part of what influenced Donald Rumsfeld, first to try to cut back severely on U.S. ground forces during his first year as Secretary of Defense for President George W. Bush, and then to insist on deploying only a small invasion force to Iraq in 2003.

This chapter begins with the issue of the so-called revolution in military affairs or RMA. It is the broadest technical subject addressed here. (In addition to helping frame subsequent discussions, it also complements the discussion of military readiness in the book's budgetary chapter.) The chapter then turns to three more specific subjects. It addresses space weaponry, missile defense, and nuclear weapons design and testing in turn. They are distinct, yet are also somewhat interrelated: space assets are quite important in missile defense, and missile defenses are designed largely to defend against enemy nuclear weapons.

This chapter is hardly a comprehensive treatment of topics in military science. Important matters such as understanding trends in miniaturization, robotics, and nanotechnology, controlling the development and spread of advanced biological pathogens, and curbing nuclear proliferation through tougher export controls are not considered. But several of the key subjects of the modern era are addressed.[2] The general approach in each of the following sections is to provide a basic overview of the relevant physics and technology issues, and then try to draw whatever policy lessons might follow.

A Revolution in Military Affairs?

In the 1990s, after the drama of Operation Desert Storm brought war to living rooms in near–real time, and displayed the remarkable effectiveness of precision weapons in modern warfare, it became popular to argue that a revolution in military affairs (RMA) was brewing. Akin to previous radical transformations in warfare, such as those brought on by the inventions of gunpowder, railroads, machine guns, tanks, and airplanes—as well as the doctrines turning those individual technologies into potent fighting forces—many hypothesized that the computer age would turn the world of warfare upside down again.[3] In the American debate, this hypothesis was advanced with breathless enthusiasm by some, who assumed that the United States would continue to lead the world in innovation and hence in new forms of warfare, but with some trepidation by others, who observed that established powers are often challenged by rising powers when new eras in warfare become possible. Various phrases have been coined to describe variants on the overall modern RMA theory, including network-centric warfare (with its associated concept of "effects-based operations" designed to attack enemy nodes of information gathering and decision-making) and fourth-generation warfare (following the previous generations of Napoleonic war, early industrial war culminating in World War I, and blitzkrieg/carrier/maneuver war exemplified in World War II).[4]

The experience of the current decade has put these RMA debates into some perspective. The 9/11 attacks themselves, as well as the difficulties faced by American forces in Iraq, have disabused most of the idea that we could be entering into a "post-heroic" or "virtual" era of warfare.[5] By the predictions of some, such an era would be characterized by conflicts in which risking casualties was no longer quite as necessary (or politically possible), at least for an advanced superpower fighting a less advanced foe, and in which stand-off weaponry (or, increasingly, robotic weapons) linked to reconnaissance systems via lightning-fast communications networks would dominate the fighting.

Since technology by itself does not a revolution make, however (with the possible exception of nuclear weaponry), an RMA could only be exploited properly if it was catalyzed by the decisions of defense policymakers. In other words, any such revolution would need to be made, not passively received. Defense innovation is a complex process requiring not just different decisions on resource allocation but on warfighting doctrines and tactics, and the ways in which different elements of military forces cooperate together.[6] To be sure that the United States and its allies benefited from the RMA, rather than being hurt by it relative to their enemies, money and the energies of key leaders would need to be shifted away from areas where trends in warfare seemed largely static and redirected to new and exciting horizons.[7] For proponents of the RMA, instructive analogies might be the eras in which wooden sailing ships, horse cavalry, or bayonet-wielding infantrymen became obsolete. Surely those who recognized these trendlines earliest were best served in future wars, since they no longer wasted scarce funds—or even more importantly in some cases, scarce hours for strategizing and training and preparing battle plans—on moribund notions of how to wage war.[8] This logic led some RMA proponents to suggest, for example, that the United States might need to preserve resources for military innovation by avoiding peacekeeping missions, reducing forward-deployed forces for deterrence in key regions of the world, scaling back its two-war combat capability, and/or canceling major weapons systems of the traditional or "legacy" variety.[9]

But how to size up this modern debate meaningfully? Asserting an RMA, and pointing to nifty new gadgetry that seems to portend or constitute major progress, does not suffice to conclude the argument. This is a classic example of a technology-centric hypothesis in which the lay observer can feel frustrated, if not shut out, by the conversation—yet one in which the nation's interests can only be advanced by combining technical analysis with broader strategic judgment. Decisions about when to wage war, whom to fight and whom not to fight, which interests to defend and which interests to recognize as indefensible (or too hard to protect at a reasonable cost) must be informed by the technical and doctrinal realities of warfare. Yet they obviously must be made only after broad consideration of many factors. As such, it is important that most non-specialists *not* be precluded from joining the conversation. Similarly, decisions about military resource allocation should be influenced by an understanding of which areas of modern military capability are potentially advancing so quickly that they merit extra attention and additional money. But those budgeting decisions also require a sense of the nation's broader priorities and interests—of which wars we are likely to need to fight, not just which wars we might most prefer, or which might best play to our current and future strengths.[10] It is important that the RMA debate not be so obtuse, arcane, or inaccessible that most policymakers and citizens feel unable to participate. It is also important that the hypotheses of RMA proponents be expressed as specifically and carefully as possible, so they can be evaluated analytically and individually.[11]

In the following pages, I suggest three ways in which this assessment can be attempted. The first is to look at history and place current technical/doctrinal trends in some perspective. Are we truly at the cusp, or in the middle, of a transition so dramatic that it qualifies as revolutionary—meaning that change is not only impressive, but disproportionately more rapid than in past recent eras?

A second approach is to look at areas of major technology more carefully. Computers are undergoing a contemporary revolution, to be sure, but is this true for other key areas of military technology as well? And to the extent it might *not* be true, can fundamental progress in computers as well as perhaps a couple other key areas of technology nonetheless themselves drive a revolution at a time when many key material underpinnings of modern militaries may not be progressing so fast?

A third approach is to try to understand outcomes in recent military battles as clearly as possible. This should then help us understand the degree to which modern technology is the driver—and the degree to which expected future progress in technology may so radicalize the way humans battle each other that we must discard many old notions of conflict to prepare for a whole new future way of war.

The Computer Age in Historical Perspective

Beyond a doubt, a computer-driven revolution is occurring in modern times, with implications that go far beyond how fast we can call up data on the Internet. Microprocessors combined with modern communications technologies such as fiber optic cable and satellite constellations guide the performance of many mechanical systems, permit creation of real-time data networks accessible not only at desks but by phones and BlackBerrys and other devices, and point the way to a pending age of robotics. These changes are remarkable. Certainly they would seem, at first blush, as dramatic as some of the technology-driven revolutions of past times, such as the invention of the crossbow centuries ago, or of gunpowder, or iron-hulled ships, or the machine gun, or blitzkrieg and aircraft carrier war. But how do we know if they portend, or ensure, a revolution in warfare?

To answer this question, we must be specific. Many historians will surely look back on this era and, with rhetorical flourish, remark on the changes that occurred. But in policy terms, the most important matter is whether changes in warfare are now so rapid, and so exponential, that they necessitate a fundamental change in how we do business.[12]

The modern American military has institutionalized change from within in the last hundred years or so. At least since the 1920s, military planners have consistently expected the next decade of war to differ greatly from the one before. And even once one moves beyond the chronology of the 1930s and 1940s, which brought the world blitzkrieg, carrier war, amphibious assault, modern radar, great strides in submarine as well as anti-submarine technologies such as sonar, and nuclear weapons, changes were rapid. The 1950s saw the coming of age of helicopters and jets; the 1960s witnessed widescale adoption of satellites and ballistic missiles; the 1970s brought huge leaps in cruise missile, stealth, infrared, and night-vision technologies; the 1980s and 1990s saw the real arrival of the modern age of precision strikes and rapid battlefield communications, facilitated in large part by modern computing, and the initial arrival of robotic technologies like unmanned aerial vehicles—followed in the current decade by the weaponization of such vehicles.

Even this cursory review reminds us that the age of computing is not without modern precedent, in terms of the remarkable new capabilities it offers. Certainly the notion of airplanes, then of airplanes flying without propellers, then of airplanes flying with rotary wings able to go straight up and down are quite remarkable. Yet the latter two of these changes rarely are viewed as revolutionary. Being able to monitor the Earth from space, or through bad weather, or at night creates a transparency to the battlefield radically different than what had been the case before—yet most of these breakthroughs in sensing technologies are often not as ballyhooed as the computer is today. Is that because the computer is so obviously more important? Perhaps, and perhaps not.

In addition, to the extent the modern era is characterized as a computer-driven period of civilian and military innovation, is it possible that the fastest rate of change may already be behind us? That is, the basic notion that we are living in an era during which the military "reconnaissance strike complex" of advanced sensors, munitions, and information networks has begun to flourish is now widely appreciated by defense planners. To be sure, that reconnaissance strike capability is constantly improving, but the core concept is now well understood and established. So even to the extent that there is revolution in the air, it is possible we are beyond the point of maximum change.[13]

These arguments are hardly conclusive; they are designed simply to be reminders of how much change has gone before us, as a way of maintaining a certain humility about the context of contemporary accomplishments. For the broader purposes of this chapter, they also provide an example of how historical perspective may be used to gain some analytical perspective on a given hypothesis or recent development. History can itself be a tool of scientific inquiry—even if it is generally not conclusive in its lessons for today.

To take the next step in evaluating the significance of modern technology innovation, and its potential for dramatically changing warfare, we need to learn more about the nature of technological progress in the world today. Is it truly most dramatic in the computer area? Are other sectors of technology changing as fast, or almost as fast, or significantly more slowly? With some provisional answers to these questions in hand, we can then return to the broader question of trying to ascertain how everything adds up—how the overall state of modern technological innovation affects what is possible for military planners today and tomorrow.

Key Areas of Technology

One report from the 1990s that was enthusiastic about the prospects of a modern revolution in military affairs argued that computers are not unique. It claimed, by contrast, that the rate of change they are experiencing *typifies* the modern era. In other words, by this logic, computers are not the outlier, they are the new norm, and most types of systems are changing comparably fast—meaning a new generation of capability emerges every few years.[14]

A more sober perspective might note that most modern airplanes, ships, and ground vehicles travel at roughly the same speed as their predecessors of twenty, thirty, or even forty years ago; that modern satellites and other sensors, while better to be sure than their predecessors, perform their jobs in roughly comparable ways; that the internal combustion engine remains the dominant power source for land warfare, ensuring huge logistical trains to support deployed armies; that small arms and explosives remain very difficult to detect from any distance, especially in complex or urban terrain.

Rather than keep the debate at such a broad level of point and counterpoint, it is better to get more concrete. Several broad categories of technology can be defined, then several subcategories within each can be further specified, and then each can be individually assessed. Such was the methodology I adopted in a book written in 2000. After a survey of the technical literature, and an appeal to basic concepts of physics that established the realm of the possible for many of them, I hazarded initial estimates about the rate of change in each area. With these approximations in hand, formulated crudely given the facts that my technical training went only through the Master's Degree level and was not dramatically improved by real-world work experience, I then traveled to numerous weapons laboratories and research centers around the country to engage in dialogue with true experts and gain their feedback. This approach is an example of how someone with less than world-class technical credentials can nonetheless wade into the scientific debate about military matters.

The spectrum of key defense technologies can be broken down in many ways. One convenient method is to focus on the following:

- Sensors
- Communications systems
- Main engine / propulsion technologies and robotics
- Explosive and kinetic ordnance, as well as new weapons technologies such as directed energy
- Defenses against weaponry

Note with this list that computers are a key element of each of the categories rather than a category unto themselves. The preceding categories directly relate to key battlefield requirements for any army—gaining information, sharing it, maneuvering, destroying the enemy, and protecting oneself.

SENSORS

The laws of physics limit progress that is being made, or that will be possible, in many areas of sensing—that is, trying to locate and identify objects of military interest through visual, infrared, radar, sonar, or other detectors. Progress in miniaturization and computing is allowing some progress in technologies like radar, while robotics is permitting reconnaissance systems to go more easily where they could go before only with great difficulty and risk. For the most part, however, progress rates are modest in sensor technology.[15]

Sonar has largely plateaued, and predictions from the 1990s that the oceans would become transparent to one form of sensor or another remain extremely far from realization.[16] (The laws of physics suggest it will take a very long time indeed to change this situation, given that most forms of radiation do not penetrate more than a few dozen meters of seawater.) Optically-based systems remain inherently limited by their difficulty seeing through walls, soil, water, and foliage. Most of these constraints apply to radar too, even if some kinds of foliage-penetrating radar show some promise. Particle beams can see a great deal, but typically only at close range, and only when substantial power sources are available. Aspirations remain to make them capable of seeing through walls or dense jungle.[17] But systems capable of such accomplishments are likely to be very expensive and probably quite large for many years.

Biological detectors are advancing, but it remains very hard to identify pathogens at any distance even when they are in aerosol form.[18] Magnetic detectors cannot easily find small arms or IEDs or other such materials except at very close range, given the number of non-military objects emitting signals themselves (and the possibility of weapons being made without metal). Chemical detectors are improving, but again need to be relatively near their quarry in order to have access to enough molecules to allow reliable detection. Systems for detecting sniper fire are being researched, but in the near future they will not prevent well-trained snipers from getting in the first shot and then fleeing.[19]

Of course, the state of the art is hardly static. For example, even if the underlying technologies on sensors are themselves advancing only modestly in most cases, the ability to proliferate sensors across more platforms, including unmanned ones, is growing. So is the ability of communications networks to share whatever information is obtained more quickly. Clever applications of such capabilities will arise, like a fledgling capability known as Ancile that identifies incoming mortar rounds, predicts their impact points, and informs individuals near those impact points of how to move to avoid harm most efficiently.[20]

But when one considers some of the bolder rhetoric of RMA proponents and compares it with the technical and physical realities, there is a case for sobriety and modesty in expectations.[21] For example, according to an Air Force document of a decade ago, sensors will soon allow us to find and identify virtually anything of military significance on the face of the

Earth in coming years. This vision is not plausible, however. Similar caution should be taken when considering, for example, the prospects of an idea put forward by the Defense Science Board in 1996—that "there is a good chance that we can achieve dramatic increases in the effectiveness of rapidly deployable forces if redesigning the ground forces around the enhanced combat cell [light, agile units with ten to twenty personnel each] proves to be robust in many environments."[22] Unmanned platforms may be proliferating, but they are still expensive to operate and hardly omnipresent on the battlefield. Communications systems may be radically better than before, but they cannot themselves generate good data and continue to rely on sensors for that data. Most of all, in complex environments such as cities, the majority of military targets remain small and well camouflaged amidst very complex backgrounds, and often shielded from most stand-off sensors by buildings or other objects. As a result, even high-tech U.S. units in such environments will often "find" their enemies by being shot at.

COMMUNICATIONS

In communications, there is no doubt that progress has been remarkable. In the last decade or so, computers have been placed on tactical fighting vehicles, data rates involving satellite communications have increased by somewhere between a factor of 10 and 100 depending on the measure used, and through frequent practice the U.S. military has greatly improved procedures to get the right real-time data to the right people on the battlefield. Underlying progress in computer technology has made all this possible by radically improving the rates at which data can be processed before being distributed, making possible what the military likes to call "Network-Centric Warfare." In Operation Desert Storm, a delay of many hours or even days arose between when one sensor on one part of the battlefield found a target and when an aircraft elsewhere could be directed to attack that target. The typical delay had shrunk to less than an hour by the end of the 1990s and to less than half an hour in recent years. The benefits of these trends are seen not only in the air and naval domains, where modest numbers of high-value platforms are the order of the day, but even in ground combat, where communications systems like FBCB2 ("Force 21 Battle Command Brigade and Below") increasingly integrate every major ground vehicle into a common battlefield picture, benefiting from GPS and other enabling technologies. Even when enemy force positions are not known, these kinds of systems help a great deal with "blue force tracking," allowing American and coalition forces to know each other's whereabouts, as evidenced in the invasion of Iraq and other recent missions.[23]

Of course, many trends in communications help potential enemies, too. Al-Qaeda and its global affiliates have learned how to communicate via the Internet very effectively, frequently changing web site electronic addresses while also varying the physical locations from which they set up and update those sites. They have learned how to avoid reliance on satellite phones and to minimize their use even of cell phones, especially in areas with strong U.S. or allied local presence. This reduces their inherent ability to coordinate quickly in some ways, but they have adapted by dispersing the technologies and the authority needed to initiate operations locally so that central headquarters are not as critical—at least not for the routine use of car and truck bombs and other such relatively simple devices. Turning to the other end of the conflict spectrum and the actions of potential nation-state rivals, countries like China are increasingly emulating the United States by constructing "reconnaissance strike complexes" of their own.

Within the communications realm at least, it is probably fair to say that trends are indeed revolutionary, even if they may help adversaries nearly as much as America and its allies. However, one fundamental vulnerability of these new communications capabilities remains: the information networks being constructed today are fragile, remarkably so in some ways. Some of the most striking vulnerabilities are in the growing use of commercial communications satellites, for example, which are not resilient to direct attack or jamming. If major powers fight each other, the resilience of their information grids will be quickly and severely tested.

ENGINES AND PROPULSION TECHNOLOGIES

Without belittling the efforts of modern engineers, if there is a single striking area of technology in which progress is not revolutionary, it is in the basics of how vehicles are powered—and fueled—on the battlefield. Visionaries about future war have talked about fast jets bouncing along the troposphere and covering intercontinental distances in two hours, or weapons in space being de-orbited to strike rapidly and precisely at targets on Earth, or ground armies quintupling their speeds while reducing fivefold or tenfold the number of forces they need to accomplish a given mission. But a careful examination shows fairly definitively that such visions are not within reach in the foreseeable future. As such, the following language used in the well-regarded 1997 National Defense Panel report is too sweeping to be accurate: "The rapid rate of new and improved technologies—a new cycle about every eighteen months—is a defining characteristic of this era of change . . ."[24] Indeed, for many areas of engine and propulsion technology, it can be debated whether there is a new cycle of technology even every eighteen years, if one focuses on the fundamentals of fuel consumption and speed.

Of course, modern jet fighters are faster than before, their engines burn at hotter temperatures, they can go further and faster on supercruise. Catamaran-hull ships can attain speeds of 50 knots or more. Solar-powered robots are being developed. Ramjets can power certain air-to-air missiles at remarkable supersonic speeds, and will get even faster as scramjet (or supersonic combustion ramjet) technology becomes available in coming years.[25] Per pound of vehicle mass, modern internal combustion engines are more efficient than their predecessors. And next-generation combat vehicles are intended to require much less fuel than Abrams tanks.

But there are striking limitations implicit in all the preceding observations and predictions. Most radically new capabilities such as hypersonic vehicles and electromagnetic rail guns relate to special-purpose vehicles that are too immature and expensive to be widely practicable in the near future.[26] Transport planes and ships as well as main warships and aircraft carriers and submarines, ballistic missiles and space launch vehicles, and battlefield trucks continue to plow along at roughly the same speeds, without radically reduced fuel requirements, relative to their predecessors of two or three or even four decades ago. Progress is measured in improvements of 10 and 25 percent from one generation of vehicle to another, not a doubling of capability and speed every eighteen to twenty-four months as with computers.[27]

If next-generation main combat vehicles require less fuel than today's big tanks and fighting vehicles, it will be largely because they will be smaller and lighter—and, inevitably, more vulnerable to direct fire, given the modest incremental rates of progress in armor.[28] (This is especially true when measured on a relative scale against the rates of progress in antitank weaponry.) The internal combustion engine of today is better than that of the 1960s, 1970s, and 1990s, but it operates not far beyond the basic parameters of such earlier vintages. Predictions like that offered by a knowledgeable observer in 1997 that the speeds of battlefield maneuver for major ground forces might increase from 40 kilometers per hour in Operation Desert Storm to 200 kilometers per hour by 2010 can now be seen to be incorrect (with today's speeds much closer to the 40 than 200).[29]

This is not to say we should abandon all modernization efforts of traditional weapons platforms. To take one example, the U.S. Air Force strongly favors the F-22 Raptor and F-35 Lightning II out of a conviction that previous generations of fighters are now less capable than a number of foreign-made combat aircraft, and that they would fare poorly against comparably trained pilots in air-to-air combat (as well as against enemy surface-to-air missile attacks). For example, the Air Force rates the Russian-made Su-30 Variant as superior to an F-15C air superiority fighter in its radar, its weapons, its electronic attack capabilities, its range, its multitarget tracking ability, and its maneuverability (in other words, all the categories it appears to consider major in importance, based on a recent briefing). Somewhat dubiously, the U.S. Air Force also gives the edge to the Chinese-made F-11B, though by a closer margin.[30] Rather than hinge everything on superior pilot training and superior networking (through systems such as AWACS control planes), therefore, the Air Force wants its future aircraft to be more capable.

To be sure, as history advances, some capabilities may become more important than others, with stealth as well as advanced sensor and fast communications systems at the top of the list. Such an argument is not necessarily a case for a categorical improvement of all combat capabilities; what matters is reducing key vulnerabilities and improving systems where advances in technology offer major benefits. For example, moving from a traditional combat aircraft such as an F-14, -15, -16, or -18 to a stealthy plane can reduce an aircraft's radar cross section (or effective reflective area for radar waves) from say 10 square meters down to 0.01 or even 0.001 square meters, based on the best available unclassified estimates—implying a tenfold reduction or more in the distance at which an aircraft can be tracked by current-generation radar.[31] These improvements can dramatically reduce aircraft attrition per sortie, by a factor even greater than the range reduction of radars trying to find and track the aircraft; they can also improve the element of surprise in an attack. These types of specific arguments based on concrete attainable improvements in existing systems should always be considered seriously by force planners. It is worth noting, however, that they are often at some odds with the RMA visionaries who tend to belittle improvements in existing weapons systems as old-fashioned as a manned combat aircraft.

Robotics do promise important new capabilities. In fact, they are already delivering them, starting with recent dramatic increases in unmanned systems in combat (including the first uses of weaponized robotics, the unmanned combat aerial vehicle).[32] Several thousand UAVs and several thousand more ground robots have been employed in the Iraq and Afghanistan wars, focused on what Jim Carafano and Andrew Gudgel of the

Heritage Foundation describe as the "Three D's," jobs that are dull, dirty, and/or dangerous.[33] Explosives ordnance disposal is a good example, and robotics allow humans to stay in the loop from close proximity, easing the challenge of making the robotics sophisticated enough to get the job done.[34] Improvements in computing and in certain mechanical technologies are making this possible. But the state of battery technology still limits what small systems can do, and big systems wind up not being dramatically cheaper or more expendable than manned systems. What robotics can do in the coming years is reduce the risk of U.S. and allied casualties—for submarines patrolling shallow waters and looking for enemy submarines or mines, for ground forces disabling ordnance or searching houses that could be booby-trapped, for aircraft hovering over enemy territory for long periods or trying to penetrate dense enemy air defenses.[35] This is a very real benefit, to be sure, but it should not be confused with radical improvements in capability, or the replacement of normal troops with automated armies.

ORDNANCE

Modern military ordnance is remarkably capable. This is not so much a statement about the explosive materials contained within them, which have evolved only modestly over the years (in terms of explosive power per unit of mass and such features), but more about their accuracy and autonomy. In recent times, every few years have brought another round of innovation—laser-guided bombs and "tank plinking" on a large scale in Desert Storm in 1991, the use of GPS-guided joint direct attack munitions (JDAMs) in the 1999 Kosovo War, the use of semi-autonomous submunitions such as sensor-fused weapons each carrying multiple SKEET warheads in Operation Iraqi Freedom in 2003. The next war could likely witness the use of fully autonomous, loitering submunitions that hover above a battlefield until a suitable target appears—such as the so-called LOCAAS (low-cost autonomous attack system), which has already been successfully tested.[36] These more accurate munitions also reduce logistics requirements for deployed forces at least modestly, by reducing the weight of typical ammunition expenditures over any given period of time.

However, it is possible that the dramatic progress in precision-guided ordnance of the 1970s, 1980s, and 1990s has actually been slowing in pace. The first of these decades brought remarkably accurate ballistic missiles, and the early use of laser-guided and infrared-guided bombs as well as the beginnings of cruise missile technology. The 1980s saw the blossoming of these capabilities, which culminated in Operation Desert Storm in 1991.

The 1990s saw the extension of these trends in precision to all-weather day-night weaponry such as the JDAM and other munitions guided by GPS satellites.

However, in the current decade, change may be somewhat less dramatic. The GPS constellations are being modernized, but less to create huge new accuracies and more to ensure dependability in the face of possible jamming. So-called ramjet technology is being applied to develop hypersonic missiles, going much faster than the speed of sound, but again the purpose is largely to protect gains realized already (by striking air defense radars and other moveable assets more quickly, so they cannot elude attack) than to create capabilities never before seen. Large-ordnance weaponry is being built, including the so-called Massive Ordnance Penetrator with 5,300 pounds of explosive (and a total weight of 30,000 pounds) and an ability to penetrate 200 feet into the soil. It could be useful against Iranian or North Korean deep underground targets, such as leadership bunkers or nuclear weapons facilities. But the number of aimpoints for which it is needed are a very small single-digit percentage of the overall target set in most conflicts, and the likelihood of having adequately reliable targeting data to use such ordnance most effectively is modest.[37]

It is important here to note that as the pace of innovation may be slowing for the United States, American competitors may be catching up. For example, in coming years China could gain the ability to use large numbers of precision submunitions launched from maneuverable ballistic missile reentry vehicles. These could, in theory, make it quite impractical to use airfields lacking hardened shelters; and even those with shelters could have their runways threatened.[38]

CONCLUSIONS AND POLICY LESSONS

So which technology trends are most important? And what do they say about the prospects for a military revolution? More importantly, what do they say about the need to reallocate resources and priorities within the Department of Defense to make sure any such revolution helps the United States and its allies rather than having said revolution catch them by surprise?

We can learn much about the answers to these questions from the real-world laboratory of the battlefield. For all the progress in international security since World War II, and the relative infrequency of wars between the major powers, there are still enough wars among regional powers, between the major powers and smaller powers, and within states, that we can see vividly what modern technologies and other new aspects of the contemporary era are doing to change the nature of combat. For students of warfare, many questions about technology issues such as the RMA hypothesis can be settled—or at least informed—by careful study of actual modern combat. Consider the following recent trends, and their apparent implications for assessing the viability of the RMA hypothesis:

- The Marine barracks bombing in Lebanon in 1983 and the Somalia debacle of 1993 portended problems seen subsequently for the United States—on September 11, 2001 and in the wars of Iraq and Afghanistan, respectively. They suggested that simple technologies on a complex urban battlefield involving irregular forces would remain—and will remain—very challenging for countries like the United States despite their huge technological advantages in many realms of weaponry.
- Operation Desert Storm, while famous for the use of precision weapons and the dawn of a new age of modern airpower, was also instructive for revealing the importance of traditional ground combat skills. American troops overwhelmingly defeated Iraqi forces as much due to their basic competence in maneuver, and the Iraqis' basic incompetence in simple tasks such as properly hiding military assets and properly employing advance guard outposts to provide warning of pending attack, as due to technology. In fact, this was not a unique characteristic of Desert Storm, but one example of a broader reality about modern warfare in which the lethality of weaponry has become so great that militaries not getting the fundamentals of warfare right open themselves up to the likelihood of rapid and overwhelming defeat.[39]
- The Kosovo War showed the promise of airpower, but it could be called a success only because Serb irregulars displaced rather than killed the Kosovar Albanians they sought to defeat, allowing NATO to coerce the Serbs over time into letting those ethnic Albanians back to their homes. Had the Serbs resorted to direct killing instead, the airpower campaign would have been too slow and indirect, and it would have been hard to declare the outcome a victory.[40]
- The successful war against the Taliban and al-Qaeda elements in Afghanistan was an impressive vindication of technologies such as satellite-guided JDAM bombs, unmanned aerial vehicles, and handheld GPS "compasses" and laser rangefinders held by U.S.

- forces and their allies. But it was only possible due to the excellent traditional skills of American special forces and CIA operatives riding horseback through rough terrain, living and working with their Afghan allies, and ferreting out the locations and identities of enemy forces. It also required that Northern Alliance Afghan forces allied with the United States engage in traditional combat at times to defeat Taliban and al-Qaeda forces who were not particularly vulnerable to standoff attack with advanced ordnance in many settings.[41]
- The Iraq invasion of 2003 was less a triumph of networked systems and high technology (though these admittedly helped enormously) than of proficient armored maneuver, close combat by American and British ground forces against Iraqi regular forces and Saddam Fedayeen, and smart adaptation to difficult battlefield conditions (like surprise flank attacks against invading forces in Iraq's southern cities, on the human side, and sandstorms on the natural side).[42]
- Most of all, the enormous challenges faced by U.S. forces in Iraq since Saddam was toppled in 2003 show how much about warfare has not changed. Finding enemies within a civilian population remains arduous work. Detecting small arms, explosives, and IEDs remotely and reliably is hard even for a country devoting billions of dollars in crash research efforts to the task. Reversing the trends and becoming more successful in Iraq in 2007 depended far more on returning to old-fashioned counterinsurgency tactics such as adequate force-to-population ratios, street patrolling with an emphasis on providing civilian population security, and effective partnering with local allies than on high-tech.

Of course, the situation is not either/or. We need not conclude definitively that we are living in revolutionary times, or not. Nuance is acceptable, and it is reasonable to conclude that there are elements of revolution and major transformation demanding our attention as well as major changes in policy, while other aspects of warfare change less quickly. The fact that technical progress has sometimes been exaggerated hardly means it is unimportant, and the fact that it seems gradual when we are living through it does not mean it is slow in historical perspective. In the Afghanistan war, for example, it often became possible to take information from a sensor and get it to a "shooter" literally within minutes. This was perhaps not the result of a clear single technical breakthrough so much as a gradual improvement in procedures over the years that complemented ongoing progress in sensor systems and communications capabilities, as well as the proliferation of sensor technologies (often on unmanned aerial vehicles) that had not previously been numerous enough to create a continuous surveillance capability.[43] But a measured view about the so-called RMA may help avoid overly faddish beliefs that new types of combat have suddenly become predominant, with potentially serious consequences for how the United States allocates defense resources and plans for war.

THE MILITARY USES OF SPACE

Space is a region from which the United States now does far more than monitor nuclear weapons and missiles. In addition to traditional reconnaissance and early-warning missions, space is now the place from which the United States coordinates its conventional wars in real time. Information on battlefield targets is sometimes acquired there; information about these targets, as well as most other data, flow through space to allow rapid, high-volume, and dependable transmissions.

Several broad questions about space policy require a level of understanding about the basic science and physics of using space for military purposes.

- Are there additional ways in which space can be productively employed for military organizations, beyond those already being exploited?
- Are weapons based in space capable of providing many capabilities that Earth-based systems cannot?
- Is the nature of the space environment amenable to arms control measures?

A Brief Primer on Space, Orbits, and Satellites

Near-Earth space is home to a wide range of military and civilian satellites, not to mention vast amounts of debris that can interfere with satellite operations. Assets in space also require assets on the ground, and links with the ground, to provide services to military users of satellites.

Satellite Orbits

Most satellites move around Earth at distances ranging from 200 kilometers to about 36,000 kilometers. This region is divided into three main bands. Low Earth Orbit (LEO) extends out to about 5,000 kilometers. Geosynchronous orbit (GEO) is the outer band for most satellites. It is 35,888 kilometers or 22,300 miles above the equator of Earth. At that altitude, a satellite's revolution around the Earth takes exactly twenty-four hours, meaning it remains over the same spot on Earth's equator continuously. Medium Earth Orbit (MEO) is essentially everything in between LEO and GEO. MEOs are concentrated between 10,000 and 20,000 kilometers above the surface of Earth.[44]

The range of LEO orbits begins just above Earth's atmosphere, which is generally considered to end at an altitude of about 100 kilometers. The altitude of LEO orbits is less than the radius of Earth (which is about 6,400 kilometers, or almost 4,000 miles). In other words, if one viewed low-altitude satellites from some distance, they would appear quite close to Earth, relative to the size of the Earth itself. The dimensions of geosynchronous orbits are large relative to the size of Earth (though they are still small relative to the distance between Earth and the moon, about 380,000 kilometers). Earth's gravitational field, together with the velocity (speed and direction of movement) of a satellite, establish the parameters for that satellite's orbit. Once these physical parameters are specified, the orbit is determined and trajectories are predictable, unless and until a maneuvering rocket is subsequently fired.

Close-in satellite orbits take as little as ninety minutes to complete a tour around the planet. As noted, geosynchronous orbits take exactly twenty-four hours. Satellites in close-in circular orbits move at nearly 8 kilometers per second; those in geosynchronous orbit move at about 3 kilometers per second. Those following intermediate orbits have intermediate speeds and periods of revolution about Earth.

Satellite orbits are generally circular, though a number are elliptical, and some are highly elliptical—passing far closer to Earth in one part of their orbit than in another. Satellites may move in polar orbits, passing directly over the North and South Poles once in every revolution around Earth. Alternatively, they may orbit continuously over the equator, as do GEO satellites, or may move along an inclined path falling somewhere between polar and equatorial orientations.

Getting satellites into orbit is, of course, a very challenging enterprise. They must be accelerated to very high speeds and properly oriented in the desired orbital trajectories. Modifying a satellite's motion is very difficult once the rocket that puts it into space has stopped burning; generally, the satellite's own small rockets are only capable of fine-tuning a trajectory, not changing it fundamentally. Even though satellites in GEO end up moving much more slowly than satellites in LEO, they must be accelerated to greater initial speeds (typically about 10.5 kilometers per second). This is because they lose a great deal of speed fighting Earth's gravity as they move from close-in altitudes to roughly 36,000 kilometers above the planet's surface. In fact, a three-stage rocket that could carry a payload of fifteen tons into LEO, for example, could only transport three tons into GEO. For that reason, it typically costs two to three times as much per pound of payload to put a satellite into GEO as into LEO.

Even getting to LEO is difficult. For example, putting a payload into Low Earth Orbit typically requires a rocket weighing 50 to 100 times as much as the payload. Consequently, even Low Earth Orbit launch is stubbornly expensive, despite longstanding efforts to reduce launch costs; putting a satellite into LEO typically costs from $3,000 to $6,000 per pound (though some Ukrainian and Chinese launch services charge less than $2,000).[45] There are some hopes that the next generation of launch vehicles will be less expensive—but probably not radically so.[46]

Most satellites weigh from 2,000 pounds to 10,000 pounds, roughly speaking, implying launch costs of about $10 million for smaller satellites in LEO to $100 million for larger satellites in GEO. Exceptions exist, however, including the large imaging satellites known as Lacrosse and KH-11, each of which is believed to weigh about 30,000 pounds. In addition, most satellites have dimensions ranging from 20 feet to 200 feet and power sources capable of generating 1,000 to 5,000 watts—though again, imaging satellites would be expected to exceed these bounds.

Current Satellites

Currently, about 17,000 items of space debris are large and visible enough to be tracked by U.S. monitoring equipment. Given the state of technology at present, that implies a diameter of at least ten centimeters (about four inches). Less than 1,000 of these objects are working satellites; the rest are old satellites or large pieces of debris from rockets.[47]

The vast majority of most countries' current satellites are in LEO or GEO. In fact, excluding Russian satellites (with their particular history and their particular circumstances, servicing a large northern country), each of those zones accounts for about 45 percent of the satellites in active use today. Another 5 percent are in MEO; most of the remainder are located in highly elliptical orbits.[48]

In many cases, the dividing line between military and civilian satellites is blurred. The United States uses GPS satellites for military and civilian purposes. It buys time on commercial satellites for military communications. The U.S. military and intelligence services often purchase imagery from private firms, especially when relatively modest-resolution images (with correspondingly larger fields of view) are adequate. And some satellites provide weather data to the military as well as to other government agencies.

In addition to satellites, a tremendous amount of manmade junk resides in space. Probably 100,000 pieces of debris larger than a marble are in orbit—those at altitudes above 1,000 kilometers will remain in orbit for centuries; those above 1,500 kilometers for millennia. Perhaps 300,000 small objects, such as chips of metal or even specks of paint, are too small to be tracked—nevertheless, if measuring at least four millimeters in size, they are large enough to do potential harm to any object they might strike, given the enormous speeds of collision implied by orbiting objects. In 1983, for example, a paint speck only 0.2 millimeters in diameter made a 4--millimeter dent in the *Challenger* space shuttle's windshield. Only two other collisions between debris and operational satellites were known to have occurred through 2001, but with debris in low orbital zones growing at the rate of about 5 percent annually, more can certainly be expected. Indeed, a small satellite at an altitude of 800 kilometers now has about a 1 percent chance annually of failure due to collision with debris. In the range below 2,000 kilometers, there is now a total of 3 million kilograms of debris (in contrast to about 200 kilograms of meteoroid mass).[49]

To illustrate some of these general considerations more vividly, consider the current American satellite fleet. Most individual types of satellites are in LEO or GEO. However, MEO is also important due to about thirty global positioning satellites now in that region.[50] These provide navigation aid to military and civilian users. Since 2000, they have provided both types of users with their positions to within about five meters.

The U.S. military operates LEO satellites for ocean reconnaissance, weather forecasting, and ground imaging. The number of White Cloud ocean reconnaissance satellites that listen for emissions from ships probably numbers about eight, at altitudes of roughly 1,000 kilometers. The United States has at least two weather satellites, known as Defense Meteorological Satellite Program systems, in polar LEOs (they also carry gravity-measurement, or geodetic, sensors).[51]

The United States also deploys probably half a dozen high-resolution imaging satellites in that LEO zone. They come in two principal types: radar imaging satellites, known as Lacrosse or Onyx systems, and optical imaging satellites, known as Keyhole systems, with the latest types designated KH-11 and KH-11 follow-on or advanced satellites. The Lacrosse radar satellites operate at roughly 600 to 700 kilometers above Earth, are capable of effective operations in all types of weather, and produce images with sufficient clarity to distinguish objects one to three meters apart. The KH satellites are capable of nighttime as well as daytime observations, by virtue of their ability to monitor infrared as well as visual frequencies. They acquire information digitally and transmit it nearly instantaneously to ground stations. Their mirrors are nearly three meters in diameter, and they move in slightly elliptical orbits ranging from about 250 kilometers at perigee (point of closest approach to Earth) to 400 kilometers or more at apogee. Ground resolutions are as good as roughly fifteen centimeters (six inches) or even less under daylight conditions. They can take images about 100 miles to either side of their orbital trajectories, allowing a fairly wide field of view.[52] They do not work well through clouds, however.[53]

In GEO or near-GEOs, the United States deploys communications satellites, early-warning satellites for detecting ballistic missile launch, and signals-intelligence satellites for listening to other countries' communications or the emissions of their electronics systems, such as surface-to-air radars. For example, in the communications domain the United States has numerous Air Force packages on various hosts (including GPS satellites in MEO) for tactical communications; eight Follow-On satellites (or UFOs!) operating in the UHF frequency band that replace the Navy FLTSAT-COM satellites for naval communications; two global broadcast system satellites for transmission of video and other high-data multimedia; nine defense satellite communications system (DSCS) satellites; and five MILSTAR (Military Strategic and Tactical Relay) satellites hardened against nuclear effects and jamming for critical communications. It also has at least three defense support program (DSP) satellites for early warning of ballistic missile launches (as with most of its sensitive military satellites, exact numbers are classified; often what is publicly available is the record of launches rather than of how many satellites remain operational).[54]

The United States fields a handful of signals-intelligence satellites in GEO, though like the Lacrosse, Keyhole, White Cloud, and DSP systems, their exact number is classified. The signals-intelligence satellites have in recent times included the Magnum, with an antenna reportedly 200 meters wide for eavesdropping on communications. Jumpseat satellites, flying elongated orbits, were developed to listen into communications from northern parts of the Soviet Union.

The United States puts most military payloads into orbit from launch facilities at Cape Canaveral in Florida and Vandenberg Air Force Base in California. It also operates a half dozen smaller sites for some payloads.[55]

These satellites have permitted a radical increase in data flow rates in recent conflicts—from 200 million bits per second, already an impressive tally, in Operation Desert Storm in 1991 to more than ten times as much (2.4 gigabits per second) in Operation Iraqi Freedom in 2003.[56] With the introduction of laser communications satellites over the coming decade, this progression is expected to continue, with another tenfold or more increase in capacity likely (assuming, that is, that reliable high-speed ways to link the satellites with ground stations are found—atmospheric turbulence and weather create challenges to using many types of lasers).[57]

As for tracking objects in space, today most countries conduct space surveillance using telescopes and radar systems on the ground. Only the United States has a system providing some semblance of global coverage (though its southern hemisphere capabilities are quite limited). Its monitoring assets are located in Hawaii, Florida, Massachusetts, England, Diego Garcia, and Japan.

Consider the capabilities of a couple other countries as well. Although it has clearly fallen from its superpower status, Russia remains the world's second space power by most meaningful measures. It continues to put satellites into space at an impressive pace, averaging more than twenty-five launches a year in recent times, in contrast to a U.S. level of around forty.[58] It does so using at least eight different families of launch vehicles of many sizes and payloads, including Molniya, Soyuz, Cosmos, Shtil, and Start variants. It operates five of the world's twenty-seven major launch sites. Russia's manned space program also continues. In recent years, it has maintained a typical flight schedule of two launches with three to six cosmonauts per year.

Russia has more than forty working military satellites by recent estimates, close in quantity to the United States. They run the gamut from communications and navigation assets to early-warning satellites to electronic intelligence devices. Its satellite capabilities have been deteriorating since the dissolution of the Soviet Union, though some efforts of late have been made to restore these capabilities, for example with satellite navigation systems akin to GPS.[59]

China has more than thirty satellites in orbit and has been increasingly active, with five to ten launches per year in recent times.[60] It operates three launch sites and is an increasingly popular low-cost provider of orbiting services. It also is working on a manned space program, run by the People's Liberation Army (PLA), and put its first astronaut into space in 2003. It also hopes to put an unmanned vehicle on the moon within a few years. China uses a half-dozen space launch vehicles in the Long March series.

Most are three-stage rockets whose payloads range from 2,000 to 10,000 pounds per launch. An improved family of liquid-fueled rockets is also being developed. One variant is expected to have, among other features, the capacity to lift 24,000 pounds to LEO. China is improving its satellite and space capabilities with vigor, and is interested in developing imaging satellites based on electro-optical capabilities, synthetic aperture radar, and other technologies. Its Ziyuan imaging satellites, planned in conjunction with Brazil, would have real-time communications systems to get data to the ground quickly, as would be needed for tracking mobile military targets including ships. It is also cooperating with Russia on a number of space programs, possibly including satellite reconnaissance technology, and is making progress on electronic intelligence satellites, as well as on a rudimentary GPS-like system called Beidou.[61]

If space-related technologies could be frozen in place in their current state, the United States would be in a fortunate position. It dominates the use of outer space for military purposes today, while Russia's capabilities have declined considerably. China's assets are improving, but it probably needs better real-time information grids and perhaps an electronic signals intelligence satellite to have significant capabilities against U.S. Navy ships near Taiwan, for example.[62] The capabilities of America's other potential rivals are generally rudimentary. The United States is able to use satellites for a wide range of missions, including not only traditional reconnaissance and early-warning purposes but also prompt real-time targeting and data distribution in warfare. Although some hope to develop space-based missile defense assets someday, the present need for such capabilities is generally rather limited, and ground-based systems increasingly provide some protection, in any event (see the next section of this chapter for more). Of course, it is not possible to freeze progress in technology, nor stop the dissemination of technologies already available.

Trends and Future Opportunities

Clearly many technologies related to space are advancing rapidly. The following discussion focuses on several that seem likeliest to offer major breakthroughs in military capability in coming years.

Other technologies not discussed here in detail will improve, too, certainly, but in many cases, while the new capabilities they provide will be important, they may not offer radical changes. For example, space-based radar constellations may be larger and much more capable in the future. Even then, satellites will remain very expensive (a billion dollars a piece or more, for large systems), placing limits on how fast such capabilities can be deployed. Moreover, they will not provide capabilities that are otherwise totally absent, since aircraft (like JSTARS as well as various UAVs) can provide similar types of coverage in theaters where the United States can establish air supremacy.[63]

High-Energy Lasers

Chemically fueled lasers that could destroy their targets by heating them with continuous waves of infrared radiation are being developed today by the United States as missile defense systems, and perhaps by other countries as well (see the following discussion on missile defense). They could, however, be used against LEO satellites as well, at least in theory. This makes them relevant to a broader discussion on the military uses of space.

Such lasers are usually able to convert about 20 to 30 percent of the energy released by chemical reactions into laser power.[64] To damage a soft target like paper or human skin, a total dose of about 1 joule per square centimeter is required (a joule is a watt of power applied for a second). The type of target usually envisioned for high-energy laser weapons today, for example, the metal making up the skin of a SCUD missile, might be damaged after receiving 1,000 joules per square centimeter. By contrast, many satellites could apparently be damaged after receiving as few as ten joules per square centimeter, assuming a pulse lasting several seconds, according to a 1995 Air Force scientific advisory study. (Their trajectories are also easier to predict, further easing the challenge.) The main point is that satellites can be much easier to damage or destroy than missiles like SCUDs, meaning they could be targeted from much lengthier distances.

The current airborne laser program (simply known as ABL) enjoys two major advantages over most previous laser systems (such as the so-called MIRACL laser built in New Mexico). First, it is airborne, meaning it can fly and operate above the atmosphere's densest region and above almost all clouds. Since Earth's atmosphere interferes with most kinds of visible and near-visible light, scattering or absorbing much of it, this is a great benefit. In addition, the infrared wavelength used by the airborne laser is less affected by whatever atmosphere it does encounter (a wavelength range of 0.5 to 1.5 microns is considered ideal; the ABL operates at 1.315 microns). Each ABL is actually designed to be a system of lasers. The main beam is a high-power system for destroying an enemy missile. Other lasers of lesser power on the aircraft are designed for targeting and tracking and to measure atmospheric conditions. The ABL is designed first and foremost to work against liquid-fueled short-range missiles, such as SCUDs, in their burning or "boost" phase, though it could certainly be used against any liquid-fueled rocket with comparable effectiveness. Whether the ABL would work against solid-fuel ICBMs or not is unclear.

The ABL uses hydrogen peroxide, potassium hydroxide, chlorine gas, and water as raw ingredients. A number of modules (six on the first test aircraft, fourteen eventually) will together produce a beam with a strength of about 1 million to 2 million watts and a beam roughly the size of a basketball at a range of hundreds of kilometers. It is to operate on a modified 747 aircraft, and its maximum range against a short-range ballistic missile is estimated at up to several hundred kilometers. With a single payload of chemical fuel, it could fire about twenty shots, each lasting several seconds.[65]

Due to its basic technology, the ABL inherently constitutes a latent antisatellite capability.[66] The main issue with converting the ABL into an antisatellite (ASAT) weapon probably concerns target acquisition and tracking. At present, the ABL relies on hot rocket plumes for acquisition of the target; overhead satellites would not provide such a signature. Thus, the ABL could not track and destroy a satellite unless its tracking sensors were first cued to the satellite's location by the U.S. space surveillance system. Providing the necessary data links would require software changes and perhaps even more, but it would not require changes to the basic laser system of the ABL.

What could other countries do to exploit high-energy laser technology for space weapons applications? Consider the case of China. The Pentagon believes the Chinese may have acquired (perhaps from Russia) high-energy laser technology that could be used in antisatellite operations. Some reports indicate it has investigated atmospheric "thermal blooming," an effect caused by the passage of high-powered laser light through the atmosphere that leads to the distortion and weakening of a high-powered laser beam if not properly addressed.

In the end, however, it is doubtful that China, or for that matter any other country, could develop an airborne laser capability in the next ten to fifteen years. The juxtaposition of various technologies and the resources required for such a program are probably beyond its means; it is not totally a given that the United States will itself be successful with this technology. China may soon have the inherent ability to produce a ground-based high-energy laser like the MIRACL, should it devote the very substantial resources and time needed to make such a program work. It is not clear it could build the adaptive optics and other sophisticated features that would help concentrate its power, however. Without the latter, a ground-based system would have limited capabilities for ballistic missile defense, given atmospheric effects and the fact that Earth's curvature would prevent the laser from striking most missiles during much of their trajectory. But ASAT operations are easier to contemplate, since one can wait for a clear day and for the target to fly overhead.

What about space-based lasers? They are much further from fruition than ground-based systems or the ABL. The Pentagon acknowledges that they are probably ideas for 2020 and beyond. The U.S. space-based laser program as conceived to date would employ a different type of chemical laser that makes use of hydrogen and fluorine to create hydrogen fluoride, resulting in infrared radiation at a wavelength of 2.7 microns. That is about twice the wavelength of the airborne laser and is less suitable for use within the atmosphere. Given how strongly radiation at that wavelength is absorbed by water vapor, it would probably only penetrate down to 30,000 to 40,000 feet if directed into the atmosphere from space. But against targets in space that disadvantage clearly would not matter. The fuels are light and relatively stable, which is good for long-term storage in space. In the space-based laser (SBL), a large mirror with a diameter of at least four meters and perhaps as much as eight meters would be used to create a fine beam. The mirror would have to be extremely light. It would probably need to be furled up while being deployed, and then unfolded once in space. The laser would be about twenty meters long and weigh nearly twenty tons, according to current plans. The program's goal has been to move toward a lethal demonstration of the system in orbit by 2012, but a constellation of a dozen or more satellites providing global coverage is probably at least a decade away.[67]

Each SBL would essentially be a combination of three extremely complex technologies: the laser itself, the power source for the laser, and the equivalent of a space telescope to direct the beam. Integrating these elements may be no harder than in the airborne laser. Indeed, a space-based laser would not have to deal with any atmospheric distortion of its beam, as noted. But in other ways, the challenge associated with the SBL is much greater. It is already proving difficult to put lasers with weights of 100,000 pounds or more on aircraft; it is far harder to put them into space with rockets each capable of lifting payloads less than half that weight. Even if high-powered lasers, space telescopes, and large fuel payloads could be individually orbited, assembling them in space and making them work in that environment for the purposes of missile defense or antisatellite operations is a far more challenging proposition. These challenges may or may not prove surmountable within two decades. But absent major breakthroughs in materials or rocketry, or both, the costs of building and orbiting a constellation of space-based lasers may prove excessive, even if the concept turns out workable. Should certain new laser concepts, such as the free-electron laser or all-gas-phase iodine laser, be developed in the megawatt range by then, the construction and launch costs of the basic optics alone could still prove staggering. Costs for a constellation of two dozen laser weapons were recently estimated at $50 billion or more by the Congressional Budget Office.[68]

Launch Vehicles and Rockets

As noted earlier, space is an expensive place to operate, not only because it is remote, but because launch costs are very high. Will this remain true in the future? Many concepts of future space warfare assume much cheaper future pathways to space that may or may not be realistic.

In fact, fundamental improvements in the efficiency and cost of space launch systems have been elusive for many years now. Progress in propellants and structural materials for rockets, be they launch vehicles or ICBMs and SLBMs or interceptors, has been limited. Indeed, the theoretical maximum performance of current chemical fuels is being approached. New materials used in the structures of rockets can improve performance at the margin, but major improvements are unlikely with current technology. The evolved expendable launch vehicle (EELV) program, the major U.S. effort of late to achieve greater efficiencies and lower costs in space launch operations, will do very well to reduce costs by half. In fact, it seems more likely that it will do well to reduce costs at all.

Even more futuristic weapons are being contemplated by defense planners. For example, space-to-Earth kinetic energy attack weapons could also be of interest. The basic science of these types of vehicles is not particularly challenging. However, a dedicated program to create the appropriate types of aerodynamic vehicles would be needed, as would testing. It would be necessary either to develop objects that would fall predictably through the atmosphere without deviating from planned trajectories or burning up, or to develop an aerial vehicle that could fly to its destination once it had been decelerated. But orbiting weapons and later deorbiting them does not offer advantages in speed or cost or technological feasibility, compared, for example, with ballistic missiles. Putting the objects in space is roughly as energy-intensive as shooting them halfway around the world on a ballistic trajectory; using booster rockets to cause them to descend takes comparable lengths of time to what ballistic flight requires. Because of the huge costs of putting objects in space, it is extremely rare that weapons in space can be even remotely cost-competitive with Earth-based weapons.[69] (There are notions of building a "space elevator" to reduce such costs—but of course the elevator, dozens of miles long at a minimum, first needs to be invented, proven practical and affordable, and then built!)

Microsatellites

Progress in electronics and computers, as well as improvements in miniaturized boosters, have made possible smaller and smaller satellites in recent years. These types of devices augur a whole new era in satellite technology. Beyond benign applications in communications, scientific research, and the like, one type of application could be small stealthy space mines able to position themselves near other countries' satellites, possibly even without being noticed, awaiting commands to detonate and destroy the latter. They could also use microwaves, small lasers, or even paint to disable or destroy certain satellites. Moreover, they could be orbited only as needed, permitting countries to develop ASAT capabilities without having to place weapons in space until they wished to use them.

Most devices known as microsatellites weigh ten to one hundred kilograms; nanosatellites are smaller, weighing one to ten kilograms. In recent years, experimental picosatellites—devices weighing less than one kilogram—have been orbited. Two have been put up by the United States, and there may be others in space as well, as yet undetected. But it is microsatellites that are becoming prevalent. For example, Germany, China, and the United States have all orbited satellites weighing about seventy kilograms, Brazil has put up a satellite of about 100 kilograms, and Thailand and Surrey Satellite Technology in the United Kingdom have jointly orbited a device weighing less than fifty kilograms. Advanced microsatellite programs, designed largely for research purposes but also for activities such as communications, are under way in the United States, the United Kingdom, France, Russia, Israel, Canada, and Sweden. Other countries collaborating with private firms based in these locations include China and Thailand, as well as South Korea, Portugal, Pakistan, Chile, South Africa, Singapore, Turkey, and Malaysia.

Using microsatellites as ASATs may already be theoretically within near-term reach for a number of countries. The maneuvering capability needed to approach a larger satellite through a co-orbital technique is not sophisticated, especially if there is no time pressure to attack quickly and the microsat can approach the larger satellite gradually. In June of 2000, for example, the University of Surrey launched a five-kilogram nanosatellite built for less than $1 million on a Russian booster (that also carried a Russian navigation satellite and Chinese microsatellite). The nanosatellite then detached from the other systems and used an onboard propulsion capability to maneuver and photograph the other satellites with which it had been orbited. In early 2003, a thirty-kilogram U.S. microsat maneuvered to rendezvous with the rocket that had earlier boosted it into orbit. These microsats were already near the satellites they approached, by virtue of sharing a ride on the same booster, making their job somewhat easier. But the principle of independent propulsion and maneuvering is being established. Larger maneuvering space mines are quite likely already within the technical reach of a number of countries; smaller versions may soon be, too.

Conclusions and Policy Lessons

In summary, then, a few enduring realities about the physics and technology of space systems can be distilled, and a few general answers to the questions posed at the beginning of this section can be deduced.

- Space is a complex environment in which to work, and is difficult to master. Launch vehicles often fail; satellites remain challenging to build since they require great precision in their manufacture; and satellites in orbit remain impractical to repair.
- However, countries willing to develop expertise in space can certainly be successful with time, at least for basic technologies, and they can rent commercial assets even if they cannot field their own.
- Space is definitely already militarized, through the increasing use of satellites for real-time targeting and other such tactical warfighting purposes, through a process led by the United States in the post–Cold War era. (Before this, while the superpowers used satellites for nuclear targeting purposes and very limited communications, they did not have the capacity to create "reconnaissance strike complexes" to conduct real-time operations making heavy use of space assets.)
- Space may or may not become formally weaponized in coming years, in the sense of weapons being placed in orbit, or Earth-based weapons being developed to threaten satellites. But the latent

capacity for weaponization will grow even in the absence of decisions to pursue such a route for at least two reasons. First, missile defense systems will have an inherent capability against Low Earth Orbit satellites, which mimic ballistic missile warheads in their flight profiles to a large extent. The use of the Navy's Aegis ballistic missile defense capability, employing a Standard SM-3 Block IA missile launched from a Ticonderoga-class cruiser, to shoot down an errant satellite at an altitude of 153 miles in February of 2008 underscored this point. Only small software changes were required to make the intercept possible, according to Raytheon, the maker of the missile.[70] Similar capabilities are surely inherent in the midcourse defense system based in California and Alaska. Second, as microsatellites proliferate and become more advanced and maneuverable, their inherent ability to become effective antisatellite weapons (by having their scientific payloads replaced by ordnance) will grow.

- These realities make many types of space arms control unverifiable, though it may still be feasible to place limits on certain kinds of *actions or activities* (such as debris-causing explosions and collisions) if not actual capabilities.
- LEO satellites in particular are also vulnerable to nuclear detonations. Nuclear bursts can destroy satellites at hundreds of kilometers from a detonation point, either from the blast or x-rays, and some satellites might even be affected at distances of 20,000 to 30,000 kilometers by a large blast, assuming limited shielding of the satellite.[71] In addition, a LEO nuclear burst could leave low-altitude space inhospitable to satellites (through "pumping" of the Van Allen radiation belts) for an extended period. Unhardened LEO satellites with expected lifetimes of five to fifteen years might last only a few months or less under such conditions.[72]
- Space is an expensive place in which to operate, with huge inefficiencies in using chemically fueled rockets to place objects in orbit, thus limiting the advantages of space for most military applications.
- Yet certain military functions, notably in regard to communications and reconnaissance, are efficiently conducted from space despite the burdens of placing objects in orbit. Even though costs are high, benefits are greater still. Early-warning and communications satellites in geosynchronous orbit, and GPS satellites in MEO, are additional cases in point.

MISSILE DEFENSE

Missile defense was among the most polarizing and contentious issues in American defense policy for at least two decades until the Bush administration withdrew from the ABM Treaty and proceeded to deploy ballistic missile defense systems—most notably, one for intercepting long-range warheads in the midcourse of their flight that had been developed largely by the Clinton administration. As such, while the Bush administration was eager to withdraw from a treaty that the Clinton administration had had very mixed views about, the decision to deploy had a certain bipartisan quality at some level (even as most Democrats objected to the way in which the Bush administration withdrew unilaterally and rather abruptly from the ABM Treaty). The wars of recent years then tended to keep the focus off missile defense, reinforcing the new tendency to relegate it to somewhat secondary status as a prominent issue.

That said, the issues with missile defense remain very important. The overall program is very expensive, averaging about $12 billion a year all told (including defenses against shorter-range missiles) during the Bush years. The request for 2009 was for $13 billion and the longer-term plan forecast spending of $62.5 billion over the following five years.[73] Among other things, this sum of money is to purchase twenty more midcourse interceptors for the Alaska/California system, 211 Standard Missile interceptors for the Aegis Navy system, ninety-six land-based THAAD interceptors, about 400 additional land-based and shorter-range Patriot missiles, and ten interceptors for the midcourse system to be based in Europe.[74] Missile defense remains a source of substantial contention with Russia, most acutely in regard to a possible European site for a defense base but also more generally as a symbol of unchecked American power (dating back to the withdrawal from the ABM Treaty and, in fact, the entire legacy of the Strategic Defense Initiative of Ronald Reagan). It also causes concerns in Beijing, a major power with a much smaller nuclear arsenal than Russia that could in theory be countered to some extent by American missile defenses—and also a power that could conceivably wind up in a serious crisis with America over the matter of Taiwan (even if that seems less likely at the moment). One need not oppose missile defense categorically, wish for a restoration of the ABM Treaty, or sympathize with any and all criticisms of missile defense by foreign governments to recognize the sensitivities of the issue.

Several programs are at the core of the current U.S. missile defense effort. They include the Patriot missile (for ground-based defense against missiles in the final or "terminal" stage of flight), THAAD (ground-based defense against midcourse threats of modest range), the Alaska/California system (ground-based defense, with help from a sea-based radar, against long-range missile threats), the Aegis Navy system (against missile threats over or near the sea), and the airborne laser as well as the kinetic energy interceptor (both designed to work against missiles in their "boost phase," just after launch and while they are still burning). In addition, many of these specific systems are being linked together, and fed information, by various command and control systems, radar programs (upgrades to existing radars and deployment of new ones), and the planned launch of a major satellite constellation to track warheads (and try to identify them if disguised within clouds of decoys or other countermeasures). Each of these various types of capabilities is being upgraded sequentially.

As of the end of 2008, the Missile Defense Agency had upgraded radars on land in Japan, the United Kingdom, Alaska, and California, and built a sea-based mobile radar homeported in Alaska. It had increased its tally of midcourse interceptors based in California and Alaska to thirty. It now has eighteen Aegis-class ships with the capability to intercept medium-range missiles, and a total of thirty-four SM-3 interceptors on them. And it has conducted thirty-five successful "hit to kill" intercepts in forty-three attempts, with various degrees of realism in those tests, but a clear track record of improved capability.[75]

To evaluate these plans and consider various options for missile defense, it is important to have a clear mental picture of how ballistic missiles and the technologies designed to counter them actually function. Missile defense is very hard, and given the fact that it must work with extremely high overall reliability against nuclear-tipped missiles to offer acceptable levels of protection, the advantage clearly goes to the attacker over the defender. But if the defender has a major technological advantage, and enough resources, it may be increasingly possible to neutralize some of the plausible threats a small extremist state may pose. That mixed message is where the following basic technological discussion would seem to lead.

Basic Elements of Missile Defense

Ballistic missiles are rockets designed to accelerate to fast enough speeds that they can fly relatively long distances before falling back to earth. They are first accelerated by the combustion of some type of fuel, after which they simply follow an unpowered—or ballistic—trajectory. They consist, most basically, of rocket engines, fuel chambers, guidance systems, and warheads, though the specifics vary a great deal depending on the range and sophistication of the missile.

For shorter-range missiles, the entire weapons system is generally simple. The missile usually consists of a single-stage rocket, which fires until its fuel is exhausted or shut off by a flight-control computer and then ceases functioning for the duration of the flight. The missile body and warhead often never separate from each other, flying a full trajectory as a large single object.

For longer-range missiles or rockets, the system consists of two or three stages, or separate booster rockets, each with its own fuel and rocket engines. The rationale for this staging is to improve efficiency and thereby maximize the speed of the reentry vehicle or vehicles. Putting all the fuel for a long-range rocket in one stage would make for a very heavy fuel chamber and mean that the rocket would have to carry along a great deal of structural weight throughout the entire phase of boosted flight. That would lower the ultimate speed of the warhead or warheads, reducing their range. With staging, by contrast, much of the structural weight is discarded as fuel is consumed. That makes it possible to accelerate the payload to speeds sufficient to put it on an intercontinental trajectory. Long-range warheads must reach speeds of about 4.5 miles a second (roughly 7 kilometers a second), or almost two-thirds of the speed any object would need to escape the earth's gravitational field entirely (roughly 7 miles, or 11 kilometers, a second). To reach such speeds with existing rocket fuels, efficiency in design—including rocket staging—is essential.

On long-range rockets, warheads are designed so they can be released from the missile body during flight. Generally, warheads and any decoys are released after boosting but while the rocket is still going up—that is, in the ascent phase of flight.[76] Releasing warheads from the missile is clearly necessary if multiple warheads with multiple aim points are to be used. It is also desirable since large missile bodies are subject to extreme forces on atmospheric reentry that could throw them, and any warheads still attached to them, badly off course.

In fact, warheads do not fly free and exposed. They are instead encased within reentry vehicles. These objects provide heat shields and aerodynamic stability for the eventual return into earth's atmosphere. They protect the warheads from melting or otherwise being damaged by air upon reentry and also maximize the accuracy with which they approach their targets.

Missiles may be powered by solid fuel or liquid fuel. If liquid fuels are used, it is usually considered desirable that they be storable and not require cooling or other special treatment that would involve extensive preparation before launch. Advanced intercontinental ballistic missiles (ICBMs) can use either type of fuel; Russian SS-18s use liquid fuel, for example, whereas modern U.S. missiles employ solid fuel.[77]

Missile guidance must be exquisitely accurate. Warhead trajectories are determined by the boost phase, meaning their course is set hundreds or thousands of miles before they reach their targets. To land within a few hundred feet of a target—or even a couple of miles—requires considerable care in how long the rocket motors are fired and in what direction the rocket is steered. Generally, rockets use inertial guidance systems to measure the acceleration provided by the boosters at each and every stage of their burning. Computers then integrate those measurements to plot out a trajectory for the warheads; a feedback loop then corrects any inaccuracies in how the rockets have been firing, so that when they are shut off, the warheads' ballistic flight will take them halfway around the world and land them perhaps within a few football fields of their designated aim point.

The standard simple missile carries a single warhead. It is generally large as warheads go, but not enormous—typically weighing about as much as bombs dropped from aircraft (several hundred pounds up to perhaps a ton in weight). Rockets can also carry large numbers of bomblets instead of warheads if the weapon is not designed to cause a nuclear detonation. These can carry conventional, chemical, or biological agents in smaller packages, or submunitions, distributing their aggregate effects over a larger area than a single warhead could. They could also carry radiological payloads—basically radioactive waste, designed not to explode but to contaminate, injure, and kill indirectly.

Both warheads and bomblets can be designed to explode on impact, or when reaching a certain altitude, or after a certain amount of flight time. Bombs designed to explode at a particular altitude or after so much flight time may—or may not—detonate if they accidentally strike the ground. Much depends on the details of their design; as a rule, modern U.S. warheads would not explode under such circumstances, but simpler weapons could. This fact is relevant to certain types of missile defenses that could destroy a missile but not the warheads it carried.

Long-range missiles can also have multiple independently targetable reentry vehicles, or MIRVs. Britain, France, Russia, and the United States have developed and deployed this technology. It works in the following manner. All warheads are initially within a "bus," or vehicle-sized object that separates from the rocket's third stage at the end of powered flight. The bus has mini-booster rockets of its own, which it can use to modify its own position and speed before releasing a reentry vehicle (RV) containing a warhead (and any decoys or chaff to accompany it). It can then reposition itself before releasing another RV. Based on their minor differences in position and velocity, the warheads can then travel slightly different trajectories. Magnified by the effects of fifteen to twenty minutes of high-speed long-distance flight, these minor changes in trajectory can translate into impact points distributed throughout a "footprint" perhaps 100 by 300 miles in size.[78]

As noted, a missile bus may also carry decoys. These are objects designed to resemble warheads, thereby confusing the defense's sensors and preventing them from identifying the true warhead or slowing the defense's response time. In the vacuum of space, even extremely light decoys move at the same speed as heavy warheads if given the same initial speed; air resistance is clearly not a factor, and gravity acts equally on objects of all weights. That makes it straightforward to fool simple sensors during exoatmospheric flight. More advanced sensors that can gauge the size, shape, rotational motion, temperature, or radar reflectivity of an object may be able to distinguish warheads from decoys—unless the decoys become more sophisticated or unless the warheads are camouflaged to make them resemble decoys.

The Trajectory of a Ballistic Missile

Ballistic flight is unpowered flight within the earth's gravitational field. In other words, it corresponds to what is essentially the freefall of a fast-moving object. Once a rocket stops burning, the only forces acting on it—or any warheads or decoys released from it—are because of gravity or, upon atmospheric reentry, air resistance. That makes flight trajectories predictable and essentially parabolic with respect to the earth's surface. But the other details of the trajectories vary greatly and depend on the speed of the rocket when its boosters stop firing, as well as the angle at which the rocket is pointed.

The first, or boost, phase of a ballistic-missile trajectory is a powered flight typically lasting one to five minutes. This boost phase generally lasts about a fifth of a missile's total flight time.

For shorter-range missiles, the boost phase occurs entirely within the earth's atmosphere; for long-range missiles, it generally extends beyond the atmosphere into space. Either way, during boost phase, the missile gains an upward as well as an outward or horizontal component to its velocity. For a long-range ICBM, the missile will usually be about 200 to 500 miles downrange of its launch point and have reached an altitude of about 125 to 400 miles at the end of its boost phase.[79]

Once boost phase is complete, the remainder of the upward flight is often termed the ascent phase. Upward flight ends at the trajectory's apogee, or highest point above the earth. The missile then begins to accelerate back to earth in its descent phase.

For existing ICBMs, the ascent phase begins outside the atmosphere. It would be possible for a sophisticated country to build a fast-burn missile that would complete its boost phase within the atmosphere, but that has not yet been accomplished.[80] (The atmosphere is generally considered to end at roughly sixty miles or one hundred kilometers above the Earth's surface—even though there is no true cutoff but instead an exponential decline, and some air molecules are found even above one hundred miles.)

During exoatmospheric flight, the horizontal element of the velocity of the missile and any warheads or decoys remains constant. The vertical component of velocity is reduced by gravity, eventually slowing to zero and then reversing as the missile and any objects it has released return to earth. The result is, as noted, essentially a parabolic trajectory, as the missile continues in a generally upward motion until gravity turns its trajectory first flat and then downward.

Finally, the missile and any objects it releases, including warheads, bomblets, and decoys, reenter the atmosphere—assuming they reached a high enough altitude to have left it in the first place. Typically, missiles with ranges of 300 miles (about 500 kilometers) or more leave the atmosphere; those with shorter ranges do not.

Missile bodies, warheads, and decoys slow down during reentry because of air resistance, and do so in a manner that depends on their weight, size, and shape. As a result of this air resistance, descending objects heat up. They are also subject to strong forces that may damage them structurally if they are not well built.[81]

Missiles may be flown on several different types of trajectories to cover a given distance. A missile that flies a minimum-energy trajectory will travel the maximum distance given the speed at which its rocket burns out. But missiles may also fly on what are known as lofted or depressed trajectories for certain purposes. These names are fairly self-explanatory. Lofted trajectories are those on which the rocket's flight attains a higher altitude than a minimum-energy trajectory for the same horizontal range. Depressed trajectories, by contrast, stay closer to the earth's surface than is normal for long-range flight. Both require greater speed, and hence more fuel, to cover the same distance relative to the Earth's surface.

Basic Types of Missile Defenses

One can categorize defenses by considering the range of the defensive weapons as well as the type of mechanism used to destroy a warhead. One can also distinguish defenses according to where they are based—on land or sea, in the air or in space.[82]

To date, many defense systems (such as early versions of the Patriot) have employed traditional explosives to destroy incoming warheads. But modern systems such as the Alaska/California national missile defense system of the United States increasingly use "hit to kill" technology in which high-speed collisions between interceptor and warhead destroy the latter. (Given the typical relative speeds of well over 10 kilometers per second of these objects when they approach each other, any contact virtually guarantees the annihilation of both.)

Most missile defenses to date work in a fairly straightforward and similar fashion—and in a manner not so different from the way a radar-guided surface-to-air missile works against an airplane. First, a defense battery is "told" of a missile launch, usually by communication from an early-warning satellite that senses the heat or infrared signal from the offensive missile's booster rockets. The defense battery's radar then begins to scan the sky looking for the incoming threat. Once it locates and begins to track the threat, and the incoming object is at the proper distance, an interceptor missile is launched. Its trajectory is chosen to put it in the right place to meet the incoming threat; a computer linked to the radar makes the necessary computation.

For older systems employing explosive kill methods, the approach is then typically as follows. After interceptor launch, the defense battery radar does double duty, tracking the incoming threat and the outgoing defensive interceptor missile. The interceptor missile may have a radar receiver that allows it to pick up radar echoes from the target. (Placing a radar receiver on the interceptor missile allows for more precise tracking; it is referred to as semiactive homing.) At the proper moment, a ground control station sends a radio signal to the interceptor, causing it to detonate a conventional-explosive warhead. The explosion then creates shrapnel that, if sufficiently close to the incoming warhead, should destroy that warhead. This is the basic way the Patriot missile defense system known as the Patriot PAC-2 functions.[83]

With hit-to-kill interceptors, such as the most advanced version of the U.S. Patriot system (PAC-3), the Army's theater high-altitude area defense

(THAAD), and the Navy's theater-wide (NTW) programs, the final approach is different. Equipped with many miniature boosters, they are intended to maneuver so well that they can collide directly with incoming threats, obviating the need for (and weight of) explosives. They generally also will use either their own radar (as with the Patriot PAC-3) or advanced infrared sensors (THAAD and NTW, as well as the Alaska/California long-range system) for the final homing, having first been steered to the general vicinity of a target by radar.

The Alaska/California system mentioned earlier is an example of what is sometimes called a midcourse missile defense against long-range ICBM or SLBM warheads. Such systems generally have fifteen to twenty minutes to work against ICBMs, which is one of their appeals. During that time, interceptor missiles could travel thousands of miles, meaning that, in theory, it is practical to defend an entire land mass such as the United States with a single base or two of missiles.

The interceptors could be fired as soon as an enemy launch was noticed by an infrared-detection satellite. More likely, they would be launched after radar picked up the missile following a few minutes of flight. The United States presently has radars for such purposes on its own continental coasts, in Alaska, in England, and in Greenland. These types of radars have long wavelengths that are optimal for long-range detection. A different type of radar, generally using shorter wavelengths and thus having less range but more accuracy, would then track the threatening objects. It would guide interceptors toward targets until the interceptors were close enough to pick up the threats with their own sensors. In the final approach, such sensors would provide much more accurate readings of the location of the threats than distant radars could.[84]

Several interceptors might be launched more or less simultaneously at a single threat, to account for the possibility of random failures. Alternatively, if time were sufficient, a first interceptor could be launched, and then a second or third would be launched if previous efforts had failed. This latter technique is called a "shoot-look-shoot" defense.

In fact, it could take four or five interceptors to reliably shoot down a single warhead, not only for midcourse NMD but for most types of missile defense using interceptor rockets. That is why the Clinton administration advertised its proposed one-hundred-interceptor system as capable of destroying only a couple dozen warheads. Several problems could cause a given interceptor to miss. Rocket boosters can fail—for example, during the cold war, superpower ICBMs were generally considered to have no more than 80 to 85 percent reliability.[85] Or the so-called kill vehicle could miss its target, because of random error, a manufacturing defect, or some other cause. Even if the overall interceptor reliability were as high as 80 percent, very high reliability is needed against a nuclear weapon. To obtain 99 percent confidence of a successful intercept, in this example, three interceptors would be needed per warhead. Even more might be required if several interceptors could fail for the same reason (that is, if their probabilities of failure were not simply random, and independent from each other, but linked and systemic). Since there is not a great deal of time in which to intercept warheads, moreover, it might be impractical to attempt one intercept before firing a second and third and perhaps a fourth and fifth interceptor just in case they were needed. In other words, "shoot-look-shoot" defensive tactics may not be possible, necessitating a launch of several interceptors at once against a given warhead.

Boost-phase defenses have an appeal that midcourse systems do not: they can in theory destroy a rocket before it releases multiple warheads, as well as any decoys designed to fool a defense. An example of boost-phase systems includes the airborne laser or ABL system now in development. A major difficulty with boost-phase defenses, however, is that they must be based near the enemy missile launch point. That could be on land, at sea, or in the air—but it would need to be near the enemy missile launch points in any case. Since the boost phase lasts only three to five minutes (or less for shorter-range missiles), an interceptor does not have much time and cannot cover much distance. As a result, it must begin its flight near its target. This problem is not serious if the potential missile threat comes only from small countries that border U.S. allies or international waterways. But it makes a boost-phase defense generally impractical against missiles launched from countries with large land masses, like Russia or China.

What if a boost-phase defense were based in a low orbit in space? Even then, a space-based interceptor would need to be in the right place at the time a missile was launched, since it would not have much time to complete the intercept before the offensive booster stopped burning. So the defender would need to put interceptors in many different orbits, spacing them appropriately (the interceptors would be in constant motion relative to the Earth's surface). A simple calculation shows that only one out of several dozen interceptors might be, by chance, in the right place at the right time to intercept a given ICBM. So even to have the capacity to intercept five to ten enemy missiles, several hundred interceptors could be needed.[86]

Looking at the geometry and geography of various boost-phase options, as well as existing technologies and the likely growth in cost over time of key missile defense components, the Congressional Budget Office estimated costs for a number of modestly sized boost phase missile defenses. Its 2004 study considered five options. The first three involved land-based or sea-based interceptors of varying speeds, assumed that sixty would be needed at a total of ten sites (for any one of the options), and estimated investment costs at up to roughly $15 billion to $30 billion. Operating costs would add another $11 billion or so over twenty years. Space-based options would involve roughly 150 to 350 interceptors, respectively (since the Earth's rotation below orbiting interceptors would guarantee that most of them would be out of position at any given moment). Investment costs could reach $22 billion to $35 billion, with estimated twenty-year operating costs adding from $22 billion to $50 billion more.[87]

Even lasers, which produce beams traveling at the speed of light, would need to be located near missile launch points. Otherwise, their beams would be too weakened by the atmosphere, or by the inevitable spreading of a light beam that occurs over distance (known as diffraction) even in the vacuum of space. The beams could also simply be blocked by the earth's curvature.

Missile defense systems would generally be alerted about the launch of an enemy missile by infrared-detection satellites high above the earth. The satellites would see the strong heat signature of the rocket. Although such signals have occasionally been confused with forest fires and other hot emissions from our planet over the years, the combination of experience, more sensitive satellites, and better computers makes such confusion less likely all the time. As noted earlier, U.S. early-warning satellites are "parked" in geosynchronous orbit about 22,000 miles (or roughly 36,000 kilometers) above the Earth's surface. At that height, an object orbiting the Earth completes a full revolution once every twenty-four hours—the same speed at which the Earth's surface rotates. As a result, the satellite remains above the same region of the planet continuously.

Countermeasures

Missile defense technology is surely improving. But adversaries can adjust. Even relatively unsophisticated enemies would surely do everything in their power to make a defense's job as hard as possible—and they would probably have some fairly simple ways to do so.

One approach would be to fire more missiles than the defense has interceptors, simply saturating the defense and ensuring that some offensive weapons could not be intercepted. If the attacker had MIRV technology, saturating a midcourse or terminal defense would be even easier and require even fewer missiles.

Against defenses that can only work outside the atmosphere, in the vacuum of space, an attacker could choose to fly its shorter-range missiles on trajectories that would never leave the atmosphere. Some defenses only work in outer space (or in the very high parts of the atmosphere) because they depend on sensitive infrared detectors to home in on a target—and such detectors can be blinded by the heat generated by air resistance, particularly if an interceptor missile is traveling at high speed. Keeping trajectories within the atmosphere would require an attacker to shorten the range of many of its missiles. But for many scenarios that would not be a steep price for an attacker to pay. Rather than flying its missiles on depressed trajectories, an attacker might also move its missiles as close as possible to their target (for example, Chinese missiles aimed at Taiwan could be placed near the Taiwan Strait before launch, as indeed they have been by Beijing). In that case, their natural trajectories would be lower and their durations of flight would be reduced—preventing some defenses from having enough time to intercept them.

Against any defense that must work in the vacuum of outer space, the attacker has its greatest range of options.[88] In this exoatmospheric or midcourse region, a warhead would generally have separated from its missile—or could be designed to do so almost immediately after boosting was complete. (As noted, an advanced country could design even its long-range missiles to complete their boosting while within the atmosphere, though a less sophisticated country might not be able to.)[89]

Outside the atmosphere, air resistance will not separate out the generally lighter decoys from the heavier warheads (as it would do for the Patriot and other TMD systems that operate within the atmosphere).[90] In outer space, even extremely light decoys would fly the same trajectory as true warheads, so speed could not be used to distinguish the real from the fake. To mimic the infrared heat signature of a warhead, thereby fooling sensors that measure temperature, decoys could be equipped with small heat generators, perhaps weighing only a pound. To fool radars or imaging infrared sensors, warheads and decoys alike could be placed inside radar-reflective balloons that would make it impossible to see their interiors.[91] Decoys could also be spun by small motors so the balloons surrounding them rotated at the same speed as real warheads, in case the defense's radar was sensitive enough to pick up such motion.

There is some chance that lighter decoys could be distinguished from heavier warheads based on how they moved away from the bus. If pushed away by something like springs, lighter decoys would tend to move faster than heavy warheads, assuming springs of similar force. But detecting such differences in motion would require extremely precise sensors. The attacker might also compensate by issuing chaff just prior to releasing decoys and warheads—to prevent radars from seeing what happens during the release. It is for these reasons that the decoy problem is acute, and possibly not solvable for the foreseeable future, in the case of midcourse defenses.

Decoys like those mentioned here are not trivial to make, however—and might work only if repeatedly flight tested. (A test of a missile defense system by the United States in late 2008 involved decoys that failed to function properly.)[92] Balloons need to be inflated in outer space. Some type of mechanism must physically separate each decoy from its host vehicle as well—something that is easy to do for Russia (or the United States, Britain, or France) and others that have mastered MIRV technology, but a bit harder for countries that have not. (Most states that are of concern to the United States are highly unlikely to have MIRV technology anytime soon.) The associated technology is fairly simple, but making it work in the laboratory is not the same as making it work at high speed in outer space, especially after a high-acceleration trajectory through the Earth's atmosphere. (It bears re-emphasis that in late 2008 the United States conducted a test in which decoys failed to deploy properly.)[93]

Making decoys work within the atmosphere is even harder. It can be done, but it requires decoys that can overcome the effects of air resistance so as not to slow down more quickly than real warheads would. Decoys that could mimic warheads within the atmosphere therefore might need small booster rockets. Alternatively, they could be made small and dense, so they would fly the same trajectories as heavier but larger warheads (since the rate of slowing from air resistance increases with an object's size as well as its weight), though in that case their radar signatures might give them away.

Against boost-phase defenses, countermeasures are also possible, though they are relatively difficult to make. As noted, boost phases could theoretically be shortened to minimize the time a defense would have to home in on the hot rocket booster. Against interceptors that would track a rocket's plume, contaminants could be put in the rocket fuel to make its plume asymmetric and potentially lead astray any interceptors that might home in on the midpoint of the plume (unless the interceptors also had an additional sensor). Against lasers, a rocket could be rotated, or given a shiny external surface that would reflect most incoming light. Finally, rockets could also be launched from remote locations on cloudy days when infrared detection satellites might not detect their heat signatures immediately—reducing the time when boost-phase defenses could work.

In short, the missile defense job involves not only very advanced technologies but a complex interaction between offense and defense. Moreover, the tools available to each side are different, and in many cases advantageous to an attacker, meaning that even a less sophisticated attacker may be able to compete successfully with a technologically advanced defender. The broad message here is that one must ask about the likely offensive countermeasures that could be deployed against each and every different type of defense. Missile defense is not pure science; it is an interactive, competitive, action-reaction process.

Conclusions and Policy Lessons

Missile defenses are improving, but the task is inherently very challenging. In addition, "the enemy gets a vote," and can employ various countermeasures to challenge a defense that may have performed well against a single easily distinguishable warhead in a simulation or test. On balance, the offense is in a stronger inherent position than the defense, especially when nuclear weapons are involved (since they require any meaningful defense to have a very high probability of successful intercept). However, it is worth bearing in mind as well that countermeasures are not trivial to perfect, especially for countries with small warhead and missile inventories and limited military resources or diplomatic "space" within which to test. The following points provide some additional detail to substantiate these broad conclusions.

- Existing rocket capabilities, when juxtaposed with greater computing power and better sensors, offer meaningful new options for missile defense. The advent of hit-to-kill technology manifests a new accuracy and quickness in sensors, computing, and resulting course adjustment for small "kill vehicles." For example, the midcourse system for missile defense, begun under the Clinton administration, deploys four small "divert thrusters" on its 140-- pound exoatmospheric kill vehicle (EKV). It has already struck its target on several occasions, revealing the remarkable quickness and precision of a device that is trying to "hit a bullet with a bullet" (notwithstanding other potential limitations in the system due largely to the likely effects of enemy decoys).

- These tests were not entirely realistic, even in simulating intercepts against simple threats not including decoys, and numerous tests failed. So a mixed message arises from these trends: Remarkable new things have become possible in the realm of missile defense, but to work effectively many things must go right, including coordination between sensor systems, central headquarters, and interceptor bases typically spaced thousands of kilometers from each other.[94]
- Countermeasures such as Mylar balloons released in outer space by a missile (along with the actual warhead or warheads) could defeat a system such as the earlier-noted midcourse defense system of the United States (now featuring interceptor missiles based in Alaska and California, with plans to deploy similar capabilities in Poland and the Czech Republic). They can mimic the signature of actual warheads beyond the capacity of current sensors to distinguish real warheads from decoys.
- Releasing decoys (and having balloons properly inflate in space), however, requires some of the technologies used in building MIRV'ed rockets in the Cold War that were not trivial even for the superpowers to develop. Moreover, a country like North Korea may not have the political or diplomatic ability (or the excess rockets) needed to do enough tests to verify that its procedures for releasing decoys really work.
- Boost-phase defenses that would destroy a rocket before it could release any warheads and decoys can overcome these countermeasures. But they may be vulnerable to certain other countermeasures. Most of all, they must generally be based near the potential adversary in advance of a conflict, given the short amounts of time available to do boost-phase intercepts.
- Someday, space-based missile defenses may become possible in theory, providing an answer to the difficulties of having boost-phase defenses predeployed in the right places before a conflict. Advances in processing power and miniaturization could also make a concept like "brilliant pebbles" of Reagan-era "Star Wars" fame more feasible than in the past. The idea is to base small interceptors in space for ballistic missile defense, igniting their boosters when necessary to attack a ballistic missile or its warheads. The Missile Defense Agency is hopeful that a concept for a boost-phase interceptor can be developed within half a dozen

years, initially using ground-based rockets but perhaps shortly thereafter space-based interceptors as well.
- Making a single brilliant pebble technically feasible, however, is a far cry from populating low Earth orbit with enough of them to provide even a limited national missile defense capability. Because such pebbles would always be in motion relative to Earth, and because only a pebble that was near a ballistic missile at the time of launch could destroy it, given the short timelines available for intercept, at least several dozen pebbles would be needed in orbit for every missile that might need to be destroyed. In addition, the brilliant pebbles would not remain in orbit indefinitely. They would probably have to be replaced every ten years or so, necessitating an average, ongoing annual investment of at least several billion dollars (even for a very limited system), given current concepts for how to build such a missile defense constellation.[95]

Nuclear Weapons, Nuclear Testing, and Nuclear Proliferation

A technical issue of great importance is nuclear testing. It is relevant to maintaining deterrence for the United States, but perhaps even more importantly in the modern era, it is of importance in addressing the nuclear nonproliferation agenda. To put it directly, if testing can be impeded or stopped by international accord and resulting international pressure on any would-be violators, can nuclear proliferation be slowed? To get at such questions, this section begins with a primer on how nuclear weapons work.

Basics of Nuclear Bombs

Fission bombs, the simplest type, use either enriched uranium (U-235), or reprocessed plutonium (with Pu-239 the key isotope) created in a nuclear reactor, as their core material. Either one is capable of undergoing a chain reaction, meaning that once some atoms within a given mass of material begin to split or fission, they can cause an exponentially increasing number of atoms to themselves fission, in a process that accelerates extremely fast. Neutrons produced by the process of one atom splitting are sufficient in number, and typically endowed with the right amount of energy, that they can, on average, cause more than one atom to split themselves, assuming a sufficiently large amount of material is present in a condensed space. Hence, a chain reaction occurs, with one fission resulting from another, in a process that builds upon itself. The point here is that, with *more than* one fission resulting on average from any previous one, the process escalates exponentially. The rate at which this occurs is fast enough that, if the materials are appropriately sized and shaped, they can create enormous numbers of fissions—and enormous energy—before the weapon blows itself apart.

A number of things must go right for this process to work. The correct amount of material must be present, and the weapon must be constructed in such a way that it does not destroy itself before a large yield is created.

One way to build a bomb is with enriched uranium 235—the scarcer isotope of the two found most in nature (the other being U-238, which makes up 99.3 percent of natural uranium). Since U-238 is not very prone to fissioning, even in the presence of lots of free neutrons, bombs using a uranium chain reaction as their source of explosive energy cannot be built unless the concentration of U-235 is greatly increased through a process like centrifuge rotation or gaseous diffusion. With these methods, the slightly lighter weight of U-235 means that on average molecules containing it have greater speed than those with U-238, so through a repetitive process of enrichment the concentration of the U-235 atoms or molecules can be increased if a mechanism to separate faster molecules from slower ones is employed.

Once adequate amounts of U-235 are available, typically twenty kilograms or more, the uranium can be put into two main chunks, neither one large enough to generate a chain reaction. (This is because, if the mass is small enough, those natural fissions that do occur in the uranium generally produce neutrons that escape from the mass into space, rather than encountering and being absorbed by new uranium atoms. So the chain reaction process never gets going.) But when the two chunks are joined, as with a "gun-assembly" weapon like the Hiroshima bomb, they produce a large enough mass to "go critical." If the uranium is partially surrounded with materials that tend to reflect neutrons—so any neutrons headed for the open tend to return to the uranium mass and have a chance at causing new fissions—and if it is also surrounded with a tamper that slows down the process of the explosion, allowing more time for new chain reactions to occur—the yield can be further enhanced.

Built in this way, the Hiroshima bomb had a yield of ten kilotons (the equivalent of 10,000 tons of TNT) and destroyed a region with a radius of about one kilometer; such a weapon detonating in New York could easily kill 100,000.[96] Just to underscore the power of nuclear explosives, that much energy is created by the complete fissioning of just half a kilogram of uranium or plutonium. (In other words, despite all the efforts to reflect neutrons and slow the blast process, nuclear explosives are not very efficient, and do not typically consume the majority of their nuclear "fuel" in the course of an explosion.)[97]

For today's nuclear powers, the most common way to build a fission weapon is not to use a gun-assembly uranium weapon, but to surround a shell of plutonium with conventional explosive. The explosive must have multiple detonators so it is simultaneously detonated all around the shell, and it must be shaped correctly so the explosive force applies equally across the surface of the plutonium. When the weapon is triggered, the shell is thereby compressed, forming a sphere that attains critical mass. Less than eight kilograms of plutonium (under twenty pounds) is adequate to create such a weapon.[98]

As with uranium bombs, the yield of these weapons is enhanced considerably, for a given amount of fissile uranium or plutonium, if a neutron-reflecting material like uranium-238 or a beryllium or tungsten product is used to enclose the plutonium. Also, weapons can be made more efficient and deadly if neutron generators (designed to start the chain reaction more quickly) are used. In that way, the weapon does not depend entirely on random fissions to begin the chain reaction process, so the exponentially accelerating process can happen faster, allowing a greater yield before the weapon self-destructs. Polonium-210 is such a neutron-generating element.[99]

All of these materials, it is worth noting, are either hard to acquire or hard to work with—underscoring the degree to which building a nuclear bomb is challenging, even for some nation states, and certainly for terrorist groups. They are dangerous to handle, and they must be well machined to function correctly within a weapon. But on balance it must still be concluded that for groups able to get their hands on fissile material, the odds of it being turned into at least a crude and heavy nuclear device are fairly high.[100]

For those groups or states able to build bombs, the yield of a weapon can be increased if a weapon is "boosted" by a mixture of deuterium and tritium gas injected inside the shell of the plutonium as the weapon is detonated. That tritium absorbs neutrons and then undergoes a fusion process (atoms coming together to form new heavier atoms, rather than the opposite fission process in which a large atom splits to form more than one smaller atoms). The fusion process itself generates energy. Even more important than the energy thereby generated by fusion in the boosting process, however, are the neutrons generated by the fusion—which in turn have a high likelihood of inducing more fissions from the plutonium. Again, the goal is to maximize the number of plutonium atoms that fission quickly, before the bomb essentially blows itself up and terminates the chain reaction process.

An even more advanced bomb is the thermonuclear or hydrogen bomb. Such a bomb includes a device like that described earlier as its "primary," which gets the whole explosion going. In addition, it has a "secondary" stage, powered by x-rays from the first stage, that is designed to produce a large fraction of its energy from fusion. The yields of such thermonuclear weapons can be very large, hundreds of kilotons or even megatons (by contrast, even the most sophisticated and efficient fission bombs typically have yields limited to several dozen kilotons at most).

Some assume that in a thermonuclear or fusion bomb, the primary is a fission device, while the secondary is the fusion part of the weapon. But as noted, many primaries also employ fusion to an extent, through the boosting process. To add further to the confusion, much of the energy from the second stage of a thermonuclear bomb is typically produced by fission, since the deuterium-tritium gas is enclosed generally by a uranium-238 shell, which is good at absorbing the high-energy neutrons produced by the fusion process (even though U-238 is not good at absorbing neutrons from fissioning uranium atoms, as noted previously, given the different average speeds of those neutrons). So there is actually some fusion within the "fission" part of the bomb—the first stage—and some fission within the "fusion" part.[101]

This type of primer, while obviously not adequate for answering questions such as whether North Korea or Iran can build a nuclear weapon to fit on a missile in the coming years, nonetheless helps inform some of the questions relevant to the nuclear testing debate and specifically the issue of whether the comprehensive nuclear test ban treaty (CTBT) would enhance American security. This is considered in the following.[102]

Is a Ban on Testing Verifiable?

Large nuclear weapons detonations are easy to detect. If they occur in the atmosphere (in violation of the atmospheric test ban treaty) they are visible to satellites, and their characteristic radiation distribution makes them easy to identify. It is for such reasons that no country trying to keep its nuclear capabilities secret has tested in the atmosphere in the modern era (South Africa is the last country that may have done so). If the detonations are underground, as is more common, they are still easy to identify via seismic monitoring, provided they reach a certain size. Any weapon of kiloton power or above (the Hiroshima and Nagasaki bombs were in the ten to twenty kiloton range) can be "heard" in this way. In other words, any weapon with significant military potential tested at its full strength is very likely to be noticed. American seismic arrays are found throughout much of Eurasia's periphery, for example, and even tests elsewhere could generally be picked up. Indeed, even though it either "fizzled" or was designed to have a small yield in the first place, with a yield of about one kiloton and thus well below those of the Hiroshima and Nagasaki bombs, the October 2006 North Korean test was detected and clearly identified as a nuclear burst.[103]

The chances of detection can be reduced in only two viable ways. First, test a device well below its intended military yield, through some type of modification of the weapon's physics. (Doing this may make the device very different from the actual class of weapon it is designed to represent, meaning sophisticated extrapolation will be needed to deduce how the actual weapon would behave based on the results of the detonation of the modified device.) Second, dig out a very large underground cavity into which a weapon can be placed, thereby "decoupling" the blast from direct contact with the ground, and allowing it to weaken before it reaches surrounding soil or rock and causes the Earth to shake. This latter approach is arduous, and does not make a weapon totally undetectable. It simply changes the threshold yield at which it can be heard by American, Russian, and international seismic sensors.[104]

In summary, a country that was very sophisticated in nuclear technology might be able to do a test of a modified device that escaped international detection by virtue of having its normal yield reduced through modifications to the basic physics of the weapon. For example, less plutonium or highly enriched uranium might be used. Or if it was an advanced type of weapon, less tritium might be used. But accomplishing such engineering feats would probably be beyond the means of a fledgling power, since they are difficult even for advanced nuclear powers. (This is much of the reason why the threshold test ban treaty that limits the power of any nuclear explosion allows tests of up to 150 kilotons. It is hard to use very small explosions in the sub-kiloton range to verify the proper functioning of a sophisticated and powerful nuclear weapon.) Scientists can learn some things from artificially small explosions caused by modified devices—but probably not enough to give them high confidence that the weapon they have developed is highly reliable at its intended yield.

U.S. nuclear verification capabilities have picked up the Indian, Pakistani, and North Korean nuclear tests—even the small, relatively unsuccessful ones—in the last decade and would be able to do so with high confidence for tests from those or other countries in the future. Verification capabilities are not airtight or perfect, but their limitations are probably not solid grounds upon which to oppose a test ban treaty, on balance.

Ensuring Stockpile Reliability without Testing

Most agree that the United States needs a nuclear deterrent well into the foreseeable future. Common sense would seem to support the position that, at some point at least, testing will be needed in the future to ensure the arsenal's reliability. How can one go 10 or 20 or 50 or 100 years without a single test and still be confident that the country's nuclear weapons will work? Equally important, how can one be sure that other countries will be deterred by an American stockpile that at some point will be certified only by the experiments and tests of a generation of physicists long since retired or dead?

From the nuclear arms control point of view, some of this perception about the declining reliability of nuclear weapons might be welcomed. Declining reliability might translate into declining likelihood of the weapons ever being used and declining legitimacy for retention of a nuclear arsenal. But as a practical strategic and political matter, any test ban must still allow the United States to ensure 100 percent confidence in its nuclear deterrent into the indefinite future. Even if some uncertainty over the functioning of a certain percentage of the arsenal is tolerable, doubt about whether *any* part of it would function effectively could seriously disrupt the core logic of deterrence.

Thankfully, a reasonable confidence in the long-term viability of the American nuclear arsenal should be possible without testing. To be sure, with time the reliability of a given warhead class may decline as its components age. In a worst-case scenario, it is conceivable that one category of warheads might become flawed without our knowing it; indeed, this has happened in the past. But through a combination of monitoring, testing, and remanufacturing of the individual components, conducting sophisticated experiments (short of actual nuclear detonations) on integrated devices, and perhaps introducing a new warhead type or two of extremely conservative design into the inventory, the overall dependability of the American nuclear deterrent can remain strong. In other words, there might be a slight reduction in the overall technical capacities of the arsenal, but still no question about its ability to exact a devastating response against anyone attacking the United States or its allies with weapons of mass destruction.

Consider first the issue of monitoring a given warhead type, and periodically replacing components as needed. This is the key way the United States is maintaining its nuclear arsenal at present (its last test was in 1992). As an example, as noted earlier a typical nuclear warhead has a shell of plutonium that is compressed by a synchronized detonation of conventional explosives that surround it. Making sure the explosion is synchronized along all parts of the explosive, so the compression of plutonium is symmetrical, is critical if the warhead is to work. Over time, wires can age, detonators can age, and so forth. But these types of components can be easily replaced and their proper functioning be verified through simulations that make no use of nuclear material (and are thus allowable under a CTBT).

Things get a bit more complicated once the compression of the plutonium is considered. The interaction of the conventional explosive with the plutonium is a complex physical phenomenon that is highly dependent on not just the basic nature of the materials involved, but their shapes and their surfaces and the chemical interactions that occur where they meet. Plutonium is not a static material; it is, of course, radioactive, and ages in various ways with time. Conventional explosives age, too, meaning that warhead performance can change with time. To prevent this, in theory one can simply rebuild the conventional explosives and the plutonium shells to original specifications every twenty or thirty or fifty years, avoiding the whole issue of monitoring the aging process by simply remanufacturing the key elements of the weapon every so often. In fact, one of the fathers of the hydrogen bomb, Richard Garwin, once recommended doing exactly that.[105] But others retort that previous processes used to cast plutonium and manufacture chemical explosives have become outdated. For example, previous generations of plutonium shells (often called "pits" in the nuclear trade) were machined to achieve their final dimensions, but this produced a great deal of waste. The goal for the future has been to cast plutonium pits directly into their final shape instead (by heating the plutonium to molten form, forming it into a proper shape, and then letting it cool). Doing so, however, would create a different type of surface for the pit that might interact slightly differently with the conventional explosive relative to the previous design. And even a slight difference might be enough to throw off the proper functioning of a very sensitive, high-performance, low-error-tolerance warhead. Similarly, the way high explosives are manufactured typically changes with time. Replacing one type with another has in the past greatly affected warhead performance, even when that might not have been easily predicted based on the explosive force of the explosive—again, the detailed chemical interactions with the plutonium pit, among other such complex phenomena, are of critical importance.

So what to do? Some would argue that, for relatively small and shrinking nuclear arsenals, it is worth the modest economic cost and environmental risk (which is quite small by the standards of Cold War nuclear activities) to keep making plutonium pits and conventional explosives as we have before, even if the methods are outdated. That would ensure reliability by keeping future warheads virtually identical to those of the past. Mimicking past manufacturing processes should not be beyond the capacities of today's scientists. But this argument is not presently carrying the day, in part because of the view that there will inevitably be at least small differences in how warheads are built from one era to another even if attempts are made to avoid it.[106]

DoE has instead devoted huge sums of money to its science-based stockpile stewardship program, to understand as well as possible what happens within aging warheads and to predict the performance of those warheads once modified with slightly different materials in the future. It is a very good program, even as elements of it naturally remain debatable.[107] It is also more scientifically interesting—and thus more likely to attract good scientists into the weapons business in future years—than a program for stockpile maintenance that would do no more than rebuild weapons every few decades. But the science-based stockpile stewardship program still gives some people unease. For example, a key part of the effort is using elegant three-dimensional computer models to predict what will happen inside a warhead modified to use a new type (or amount) of chemical explosive based on computational physics. This is a very challenging process to model accurately. This method is good but perhaps not perfect.[108]

A final way to ensure confidence in the arsenal is to design a new type of warhead, or perhaps use an old design that is not currently represented in the active U.S. nuclear arsenal but that has been tested before. This approach would seek to use "conservative designs" that allow for slight errors in warhead performance and still produce a robust nuclear yield. The conservative warhead could then take its place alongside other types of warheads in the arsenal, providing an added element of confidence. Taking this approach might lead to a somewhat heavier warhead (meaning the number that could be carried on a given missile or bomber would have to be reduced), or a lower-yield warhead (meaning that a hardened Russian missile silo might not be so easily destroyed, for example). But for the purposes of post–Cold War deterrence, this approach is generally sound, and weapons designers tend to agree that very reliable warheads can be produced if performance criteria are relaxed. It could also lead to less use of toxic materials such as beryllium and safer types of conventional explosives (that are less prone to accidental detonation) than is the case for some warheads in the current arsenal.[109]

It is for such reasons that the Bush administration and Congress have shown interest in a "reliable replacement warhead" concept the last few years. To date, it is only a research concept, and a controversial one at that, with Congress not always willing to provide even research funding.[110] But it does have a certain logic and, as one element of a future American arsenal, makes sense on balance. In fact, it might even obviate the need to consider the periodic-remanufacturing idea, since it is quite clear that the United States could deploy such a warhead with extremely high confidence of its reliability.[111] Simple warhead designs are quite robust—recall, for example, that the Hiroshima bomb (a gun-assembly uranium device) was not even tested before being used.

The Case for New Nukes

Some have suggested we may need future nuclear testing for new types of warheads to accomplish new missions. For example, in the 1980s, some missile defense proponents were interested in a space-based nuclear-pumped x-ray laser. That was never particularly practical. But the idea of developing a nuclear weapon that could burrow underground *before* detonating has gained appeal—not least because countries such as North Korea and Iran are responding to America's increasingly precise conventional weaponry by hiding key weapons programs well below the planet's surface.

One possible argument for such a warhead is to increase its overall destructive depth. In theory, the United States could modify the largest nuclear weapons in its American stockpile to penetrate the Earth. This approach would roughly double the destructive reach of the most powerful weapons in the current arsenal, according to physicist Michael Levi.[112] But if an enemy can avoid weapons in the current arsenal, it could avoid the more powerful bombs by digging deeper underground. Given the quality of modern drilling equipment, that is not an overly onerous task.

Could Earth-penetrating weapons at least reduce the nuclear fallout from an explosion? They could not prevent fallout. Given limits on the hardness of materials and other basic physics, no useful nuclear weapon could penetrate the Earth far enough to keep the radioactive effects of its blast entirely below ground. But such weapons could reduce fallout. Relative to a normal bomb, it is possible to reduce the yield of an Earth-penetrating weapon tenfold while maintaining the same destructive capability against underground targets.[113] This would reduce fallout by a factor of ten as well.

That would be a meaningful change. But is it really enough to change the basic usability of a nuclear device? Such a weapon would still produce a huge amount of fallout, its use would still break the nuclear taboo, and it would still only be capable of destroying underground targets if their locations were precisely known—in which case there is a chance that conventional weapons or special forces could neutralize the site. This is the policy question that the preceding technical discussion is designed to inform, and answering it clearly requires a combination of technical and broader strategic assessments. But the technical aspects of the problem should be a part of any such calculation.

Conclusions and Policy Lessons

Nuclear weapons are complex devices that are expensive and complex to produce. That is a fortunate fact of science. More than sixty years into the nuclear age, if it had been different on the technical front, many more countries and perhaps terrorist groups could have their own fission and/or fusion weapons. But the prevalence of nuclear material worldwide, the fact that numerous countries including Pakistan and North Korea already do possess nuclear weapons, and the limitations of international controls on the movement of nuclear technologies and materials nonetheless make the present situation fraught with danger.

The focus here has been on the issues involved in building and testing a nuclear weapon. This is relevant to, among other things, the nuclear testing debate. Specifically, how much would an international accord banning nuclear tests (and punishing any violators of the regime) complicate the challenges of would-be proliferators? On the other hand, assuming nuclear weapons are still viewed as necessary for the foreseeable future, how much might a CTBT impinge on the reliability and credibility of the American nuclear deterrent? Several observations flow from the earlier discussion. Like those of other sections of this chapter, they do not tend to put policy debates to rest or resolve them definitively. Rather, they establish some boundaries to the debate and help inform the choices at hand.

- Simple nuclear weapons are not inherently difficult to make, assuming fissile materials are available and that a country has a re-

spectable engineering and science tradition that, among other things, lends itself to the precise manufacture of various components that must be built to fairly demanding specifications.
- To put it differently, mid-sized countries with good scientific and industrial capabilities can generally make nuclear weapons, while small poor countries and terrorist groups probably cannot. Unfortunately, defined in this way, the list of countries that could develop nuclear weapons would appear to include Iran and North Korea.
- Even for mid-sized countries, obtaining fissile material to build a bomb is difficult. It requires time and considerable expense, and generally involves some degree of international help and overseas technology as well.
- Even for those countries able to build simple fission bombs—something likely within the reach of most countries that can get their hands on enough U-235 or plutonium—making weapons capable of delivery by missile is hard. Given the need to be very efficient in the use of conventional explosives, casing, and other such materials (to keep weight down), it is not clear that such weapons can be developed without testing. This fact may be the chief technical argument on the positive side of the ledger when evaluating the CTBT.
- As for other aspects of such a treaty, a CTBT could be generally well verified. Only a very limited class of nuclear detonations—so small as to have quite modest military utility, especially for countries with fledgling nuclear research infrastructures—would have any realistic hope of evading detection.
- A CTBT would not weaken the essential reliability of America's arsenal. That is to say, the United States could still credibly threaten powerful nuclear retaliation against an enemy with virtually complete technical confidence.
- The detailed performance of some warheads could be cast into doubt under a CTBT, especially over a period of decades. But any reduced confidence would affect only high-performance high-yield weapons, and even then only to a limited degree in all likelihood. The detailed capabilities of America's arsenal against a wide array of targets might be reduced somewhat, but confidence in a flexible dependable deterrent force would not. Whether such possible degradation in U.S. nuclear capability would be tolerable given the requirements of deterrence is, of course, a matter for policy debate.

- A CTBT could deprive the United States of a new type of nuclear warhead, specifically one able to penetrate deeply underground. There is a chance that such a warhead could be constructed by modifying existing weapons without testing, but it is not clear. The case for weapons like Earth-penetrating warheads is in my judgment not strong; they cannot be made in a way that would prevent fallout. A broader policy judgment on their desirability requires considering the relatively modest (but nonetheless real) potential technical military advantages of any such new weapons against the harm to global nonproliferation efforts that any American resumption of nuclear testing could cause. In my view, testing to build such weapons is not worth it. That is admittedly a policy judgment informed by—but not immediately provable with—the technical assessments offered here.

Observations on the Role of Science in General Defense Analysis

Beyond the specifics of any issue, what are the broad lessons that emerge from the preceding discussions? Clearly, knowing more about scientific and technical matters in military analysis is better than knowing less; few would contest this assertion. But for the generalist, untrained in advanced science or engineering and unequipped to understand the complexities of nuclear science, space technology, military robotics, sonar and radar, lasers and radio-frequency weapons, and so on (not to mention topics not addressed here, such as advanced biological pathogens), how should one try to tackle even the rudiments of these complex subjects?

It is worth underscoring that it is definitely worth the while of any generalist to have some familiarity with the basics of military science. If nothing else, familiarity can provoke probing questions that require scientists to offer detailed answers and explanations for their views—which then can be examined for internal self-consistency and be scrutinized by other scientists.

Such a process was exemplified in the 1980s debate over "Star Wars," the project to render nuclear weapons impotent and obsolete through defenses—which was soon widely recognized, not only by scientists but the general policy community, to be an excessively ambitious goal for any technical system. (This is not to deny there were other arguments, at the time and thereafter, in support of Reagan's Strategic Defense Initiative.)

By contrast, such a process of vetting, and of understanding the fundamentals of technology, arguably did *not* happen adequately prior to the

Iraq War of 2003. At that time, Secretary of Defense Rumsfeld's hopes that a revolution in military affairs was truly underway (as well as some flawed assumptions about Iraqi politics) apparently persuaded him and other members of the Bush administration to make only minimal plans for the post-Saddam period. A greater respect for the age-old truths of military history, together with some healthy skepticism about the promise of technology and some clear-headed examination of the evidence about whether an RMA was really going to transform land warfare quickly, could have led to much more thorough preparation. And it did not take an advanced understanding of the exact capabilities of modern munitions, sensors, stealth aircraft, or vehicle and body armor to accomplish this.

By trying to grapple with basic science, a generalist can also learn important lessons about the *limits* of what rudimentary science can tell us regarding key policy debates. In other words, far from producing sophistry, a diligent effort to stay abreast of certain technological debates can remind one of which aspects of the subjects are beyond them. Learning the basics leads one to become curious about other matters, and develop working hypotheses in one's mind about them—which can then be compared with the evidence and with the arguments of more informed scholars. This is a trial-and-error process that should breed a healthy dose of humility, not overconfidence.

For example, one can learn that any major space-based laser would—since it is essentially the combination of a Hubble telescope and a high-powered directed energy device—push the limits of technology, weigh a great deal, and cost in the billion-dollar range. But a similar understanding of the basics cannot help one be sure whether or not such a weapon could actually be created within five or ten years.

Or to take another example, some understanding of nuclear physics—and of the history of the bomb—can help a generalist appreciate that producing fissile materials is typically the hardest part of making a simple nuclear weapon. Once in possession of adequate amounts of such material, most countries are likely to be able to build a workable device, perhaps with roughly the destructive power of the Hiroshima or Nagasaki bomb. And it can help one understand that, absent nuclear testing, it is much harder for any country to have confidence in a more sophisticated device, such as a thermonuclear warhead or a weapon designed efficiently and elegantly enough to fit atop a missile.

In the end, military science is far too big a part of defense analysis simply to be ignored; even nonscientists must try to wade into the subject to the best of their abilities. Thankfully, there is enough good science writing in the modern world that a diligent generalist can usually make substantial headway. The only alternative is to pretend that the scientific aspects of defense policy matters can be separated from other aspects of key decisions, and outsourced to the experts for them to resolve—which, given the interconnectedness of so many aspects of defense policy, is really no alternative at all.

QUESTION 13: What are the likely capabilities of North Korea's suspected nuclear arsenal?

ANSWER: The U.S. experience with its testing program (discussed earlier in relation to the CTBT debate) and other related considerations are of help here. No simple mathematical formula can answer the question. However, an appreciation of where the United States has itself had problems in building reliable warheads can shed some light on which technical challenges may be greatest for the DPRK.

North Korea is believed to have enough plutonium to make perhaps six to eight nuclear bombs (roughly forty to forty-five kilograms).[114] That plutonium has been, as best we can tell from remote observation (partially corroborated by eyewitness accounts), successfully reprocessed so it is separate from radioactive waste and usable in bombs. Since fission bombs are considered relatively straightforward to make, for a country able to acquire the fissile material, it has long been assumed that North Korea has had simple nuclear explosives. That hypothesis was confirmed in October of 2006 when North Korea tested a nuclear device.

The yield of that device was quite low, however, perhaps less than one kiloton. This suggests that the DPRK may not have mastered the art of compressing the plutonium shell efficiently and symmetrically, nor quickly, enough to generate a large yield. (An alternative, less likely interpretation is that it intentionally caused only a small yield, but that would imply far more sophistication than a first-time testing nation would likely possess.)

Chances are that North Korea's weapon is simple, crude, and heavy. It may or may not fit on a missile. If launched on one, it may or may not survive the g forces and heating of atmospheric reentry, given the DPRK's limited likely capacities to accurately model such environments (and the fact that even the United States has experienced challenges in these areas with its own warheads in the past).

On balance, North Korea almost surely could detonate a nuclear weapon of simple design again. But it could require a large airplane or ground vehicle to deliver any such warhead, given its likely weight and possible vulnerability to stresses and strains in a more advanced delivery process. Only further testing (and perhaps some warhead modifications) would likely give the DPRK real confidence that it could mount an advanced warhead on a missile.

QUESTION 14: With recent successful tests, has missile defense now gained the upper hand against ballistic missile threats?

ANSWER: No. Reaching this conclusion would be to go too far. To be sure, hit-to-kill technology is doing much better, given successful tests this decade of the Navy's Aegis-based Standard missile, the California/Alaska midcourse system, and shorter-range ground-based systems. It is only fair to acknowledge that these tests do rebut some past criticisms of missile defense.

But all of these tests were against simple, single, isolated targets flying predictable trajectories. None involved swarms of decoys; none involved maneuvering reentry vehicles; and none involved salvos of multiple warheads at once. At present, based on what is known about American defense systems, they likely could not handle more complex or sophisticated attacks, at least not reliably.

That said, developing offensive techniques like the use of decoys to fool defenses is perhaps harder than some assert. While not particularly complex conceptually, it takes some work and testing to learn how to dispense multiple objects from a single bus or other mechanism out in space. A clear lesson from examining the history of missile defense and space launch programs is that operations of any type in space are challenging, and unexpected mistakes often happen.

Countries such as North Korea have limited resources, as well as limited international political maneuvering room, to conduct such tests. So the DPRK's ability to develop and maintain proficiency with space launch operations may be limited. It has considerable experience with single-stage rockets, and considerable expertise in building them and deploying them successfully, but multistage rockets as well as missile bus operations are another matter.

As such, while they have significant limitations, the U.S. ballistic missile defense systems developed to date are not without some meaningful capabilities. The practical question is how much more to spend on them when many other countries could, in theory, develop countermeasures, and when it is not clear how well more advanced U.S. systems would work against such countermeasures.

QUESTION 15: Is a revolution in military affairs underway?

ANSWER: Not necessarily. To some extent, this is a matter of definition, since there is no doubt that impressive things are happening in the realm of military technology. But if the contention is that warfare is changing so dramatically as to permit a radical reconceptualization of how it will be fought in the future—and thus of how defense resources should be allocated today—the question is open.

One popular formulation of the RMA hypothesis is that we occupy a period of time when new generations of defense technology arrive every eighteen to twenty-four months. This has been true with computing capabilities, so is it true, or even partly true, more generally? Putting this thesis on the table amounts to a testable proposition that allows greater precision in how we evaluate the RMA hypothesis.

In fact, the radical changes in technology that are leading to new capabilities every two years or so result from computer advances and not much else. Moore's "law" has for decades described this pace of computing innovation. Plus, many other changes that have occurred in the modern era—such as the invention of helicopters, night vision devices, and satellite technology—were extremely impressive yet failed to be described as revolutions of the type now predicted for this computer age. While other areas of modern technology show impressive progress, notably microbiology and some robotic systems, change is occurring at only a modest pace in many other areas, like the propulsion systems, aerodynamics, and hydrodynamics of most vehicles (on the ground, in the air, on the water, and in space), as well as most types of sensor technology (sonar, radar, and optical).

The most enthusiastic theorizing about a modern RMA tended to precede the Iraq War. The latter experience has underscored the limits of technological progress, especially for those types of wars in which the United States has the most difficulty. American high-tech, while hugely useful in the war to be sure, did not produce anything close to victory for the United States in the war's early years. Only when the U.S. armed forces and their Iraqi allies reverted to time-tested (and old-fashioned, generally low-tech) counterinsurgency methods in 2007 did the tide of battle begin to turn.

The ability of specific technologies to deliver dramatic results was frustratingly slow in Iraq (and Afghanistan). For example, against improvised explosive devices, U.S. deaths remained very high despite years of concerted investment and effort. The deaths declined when the surge-based strategy started to roll up insurgents and their IED caches before the weapons could be implanted, and to a lesser extent when the United States deployed large numbers of heavy mine-resistant vehicles (a type of heavier vehicle that flew in the face of the U.S. military's goals to make most ground forces lighter and more maneuverable). Similarly, the only effective method for stopping most mortar and rocket attacks in Iraq is to prevent their launch in the first place rather than to defeat or defend against them in a more technically innovative way.

Notes

1. For a very thoughtful discussion of this issue, see Frank von Hippel, *Citizen Scientist* (New York: Touchstone, 1991), pp. xi–xv.

2. My selection of topics largely reflects my own relative strengths and limitations as an analyst; I emphasize those subjects I have studied and written on in greatest detail.

3. See, for example, Stuart E. Johnson and Martin C. Libicki, eds., *Dominant Battlespace Knowledge* (Washington, D.C.: National Defense University, 1996); Norman C. Davis, "An Information-Based Revolution in Military Affairs," in John Arquilla and David Ronfeldt, eds., *In Athena's Camp: Preparing for Conflict in the Information Age* (Santa Monica, Calif.: RAND, 1997); Joseph S. Nye, Jr., and Admiral William A. Owens, "America's Information Edge," *Foreign Affairs*, vol. 75 (March/April 1996); David A. Ochmanek and others, *To Find, and Not to Yield: How Advances in Information and Firepower Can Transform Theater Warfare* (Santa Monica, Calif.: RAND, 1998); Admiral Arthur K. Cebrowski and John J. Garstka, "Network-Centric Warfare: Its Origin and Future," *Proceedings* (U.S. Naval Institute, January 1998), pp. 29–35.

4. Albert A. Nofi, *Recent Trends in Thinking About Warfare* (Alexandria, Va.: Center for Naval Analyses, 2006), pp. 8–17, available at www.cna.org/documents/D0014875.A1.pdf [accessed April 10, 2008].

5. See, for example, Edward N. Luttwak, "A Post-Heroic Military Policy," *Foreign Affairs*, vol. 75 (July/August 1996), pp. 33–44; and Michael Ignatieff, *Virtual War: Kosovo and Beyond* (New York: Henry Holt and Co., 2000).

6. See, for example, Jonathan Shimshoni, "Technology, Military Advantage, and World War I: A Case for Military Entrepreneurship," *International Security*, vol. 15 (Winter 1990), pp. 213–15; and Robert P. Haffa, "Planning U.S. Forces to Fight Two Wars: Right Number, Wrong Forces," *Strategic Review* (Winter 1999), pp. 15–21.

7. For a good discussion of the dangers of modern trends in military technology, capability, and operations for the United States and its allies, arising

from capabilities such as precision missiles and information warfare, see Michael G. Vickers and Robert C. Martinage, *The Revolution in War* (Washington, D.C.: Center for Strategic and Budgetary Assessments, 2004).

8. For very good histories of past revolutions in military affairs, see for example Andrew Krepinevich, Jr., "Cavalry to Computer: The Pattern of Military Revolutions," *National Interest*, no. 37 (Fall 1994), pp. 31–36; Williamson Murray, "Thinking About Revolutions in Military Affairs," *Joint Forces Quarterly* (Summer 1997), pp. 69–76; Martin Van Creveld, *Technology and War: From 2000 B.C. to the Present* (New York: Free Press, 1991); Jared Diamond, *Guns, Germs, and Steel: The Fates of Human Societies* (W.W. Norton, 1997); and Max Boot, *War Made New: Technology, Warfare, and the Course of History, 1500 to Today* (New York: Gotham Books, 2006).

9. National Defense Panel, *Transforming Defense: National Security in the 21st Century* (Washington, D.C.: National Defense Panel, 1997), pp. 2, 32.

10. See Lawrence Freedman, *The Revolution in Strategic Affairs*, Adelphi Paper 318, International Institute for Strategic Studies (Oxford, England: Oxford University Press, 1998).

11. Of course, as a practical matter, many players including many civilians are typically involved in big decisions about how a nation-state prepares for, and engages in, warfare; it is critical that national leaders, civilians providing oversight to military services, civilian scientists, and members of Congress and/or parliament be as informed as possible, given their inherent roles in military debates in most countries. For historical analyses, see Barry R. Posen, *The Sources of Military Doctrine: France, Britain, and Germany between the World Wars* (Ithaca, N.Y.: Cornell University Press, 1984), pp. 41–80; Stephen Peter Rosen, *Winning the Next War* (Ithaca, N.Y.: Cornell University Press, 1991), pp. 13–18, 76–100; and Montgomery C. Meigs, *Slide Rules and Submarines* (Honolulu, Hawaii: University Press of the Pacific, 2002), pp. 211–20.

12. See Max Boot, *War Made New: Technology, Warfare, and the Course of History, 1500 to Today* (New York: Gotham Books, 2006), pp. 466–68.

13. Frederick W. Kagan, *Finding the Target: The Transformation of American Military Power* (New York: Encounter Books, 2006), pp. 393–401.

14. National Defense Panel, *Transforming Defense*, pp. 2, 32.

15. A recent Defense Science Board review study exhorts scientists to make major progress in sensors, talks about some promising trends, and notes the potential of miniaturized systems, but offers little detail about where breakthroughs seem imminent, and acknowledges that most sensor technologies will face severe limits in their range and reliability against many classes of targets. See Defense Science Board, *Defense Science Board 2006 Summer Study on 21st Century Strategic Technology Vectors, Volume II: Critical Capabilities and Enabling Technologies* (Washington, D.C.: Office of the Under Secretary of Defense for Acquisition, Technology, and Logistics, February 2007), pp. 57–65.

16. Zalmay Khalilzad and David Shlapak, with Ann Flanagan, "Overview of the Future Security Environment," in Zalmay Khalilzad and Ian O. Lesser, eds., *Sources of Conflict in the 21st Century* (Santa Monica, Calif.: RAND, 1998), pp. 35–36.

17. Richard Chait, Albert Sciarretta, John Lyons, Charles Barry, Dennis Shorts, and Duncan Long, *A Further Look at Technologies and Capabilities for Stabilization and Reconstruction Operations* (Washington, D.C.: National Defense University Center for Technology and National Security Policy, 2007), p. 52.

18. Ibid., pp. 51–52.

19. Stew Magnuson, "Technologists Take Aim at Enemy Snipers," *National Defense* (October 2007), available at www.nationaldefensemagazine.org/issues/2007/October/Technologists.htm [accessed April 11, 2008].

20. Chait, Sciarretta, Lyons, Barry, Shorts, and Long, *A Further Look at Technologies and Capabilities for Stabilization and Reconstruction Operations*, p. 50.

21. Statement of General Ronald R. Fogleman, Chief of Staff, U.S. Air Force, before the House National Security Committee, 105 Cong. 1 sess., May 22, 1997.

22. Defense Science Board 1996 Summer Study Task Force, *Tactics and Technology for 21st Century Military Superiority*, vol. 1 (Department of Defense, 1996), p. S-4.

23. Office of Force Transformation, *The Implementation of Network-Centric Warfare* (Washington, D.C.: Department of Defense, 2005), pp. 44–45, available at www.oft.osd.mil/library/library_files/document_387_NCW_Book_LowRes.pdf [accessed April 10, 2008].

24. National Defense Panel, *Transforming Defense: National Security in the 21st Century* (Washington, D.C.: National Defense Panel, 1997), pp. 7–8.

25. Dean Andreadis, "Scramjets Integrate Air and Space," *The Industrial Physicist* (August/September 2004), pp. 24–27, available at www.aip.org/tip/INPHFA/vol-10/iss-4/p24.pdf [accessed April 11, 2008].

26. Defense Science Board, *Defense Science Board 2006 Summer Study on 21st Century Strategic Technology Vectors, Volume II: Critical Capabilities and Enabling Technologies*, p. 84; and Grace Jean, "Electric Guns on Navy Ships: Not Yet on the Horizon," *National Defense* (November 2007), available at www.nationaldefensemagazine.org/issues/2007/November/ElectricGuns.htm [accessed April 10, 2008].

27. Ronald O'Rourke, *Navy Ship Propulsion Technologies: Options for Reducing Oil Use* (Washington, D.C.: Congressional Research Service, 2006), available at fas.org/sgp/crs/weapons/RL33360.pdf [accessed April 11, 2008], pp. 1–10.

28. For a concurring view, see John Lyons, Richard Chait, and Jordan Willcox, *An Assessment of the Science and Technology Predictions in the Army's STAR21 Report* (Washington, D.C.: Center for Technology and National Security Policy, National Defense University, July 2008), p. 23.

29. Robert H. Scales, Jr., "Cycles of War: Speed of Maneuver Will Be the Essential Ingredient of an Information-Age Army," *Armed Forces Journal International*, vol. 134 (July 1997), p. 38.

30. Briefing slides presented to the Air Force Reserve and National Guard Conference by Lt. Gen. David A. Deptula, Deputy Chief of Staff for Intelligence, Surveillance and Reconnaissance, U.S. Air Force, "Emerging Threats to Legacy Fighters," Andrews Air Force Base, Washington, D.C., December 5, 2007.

31. Lane Pierrot, *A Look at Tomorrow's Tactical Air Forces* (Washington, D.C.: Congressional Budget Office, January 1997), p. 77. The detection range of a radar varies with the third root of the radar cross section of an object, so reducing the radar cross section by a factor of one thousand reduces radar range by a factor of ten. See J. C. Toomay, *Radar Principles for the Non-Specialist* (Belmont, Calif.: Lifetime Learning Publications, 1982); and Merrill Skolnik, *Introduction to Radar Systems*, 2nd edition (New York: McGraw-Hill Book Company, 1980).

32. Richard Chait, Albert Sciarretta, John Lyons, Charles Barry, Dennis Shorts, and Duncan Long, "A Further Look at Technologies and Capabilities for Stabilization and Reconstruction Operations," Center for Technology and National Security Policy, National Defense University, Washington, D.C., September 2007, pp. 56–59, available at www.ndu.edu/ctnsp/publications.html [accessed July 20, 2008].

33. James Jay Carafano and Andrew Gudgel, "The Pentagon's Robots: Arming the Future," *Backgrounder No. 2093* (Washington, D.C.: Heritage Foundation, 2007), available at www.heritage.org/Research/NationalSecurity/upload/bg_2093.pdf [accessed April 10, 2008].

34. Chait, Sciarretta, Lyons, Barry, Shorts, and Long, *A Further Look at Technologies and Capabilities for Stabilization and Reconstruction Operations*, pp. 56–59.

35. For a good summary of current capabilities and trends, see Michael G. Vickers and Robert C. Martinage, *The Revolution in War* (Washington, D.C.: Center for Strategic and Budgetary Assessments, 2004), pp. 30–45.

36. Vickers and Martinage, *The Revolution in War*, pp. 14–24.

37. See, for example, "33rd GPS Satellite Launched," *Aviation Week and Space Technology*, December 24/31, 2007, p. 12; Robert Wall and Douglas Barrie, "Stealthy Strikes," *Aviation Week and Space Technology*, December 24/31, 2007, pp. 18–19; and David Bond, "Mopping Up," *Aviation Week and Space Technology*, January 7, 2008, p. 19.

38. John Stillion and David T. Orletsky, *Airbase Vulnerability to Conventional Cruise-Missile and Ballistic-Missile Attacks: Technology, Scenarios, and U.S. Air Force Responses* (Santa Monica, Calif.: RAND, 1999); and Andrew Krepinevich, Barry Watts, and Robert Work, *Meeting the Anti-Access and Area-Denial Challenge* (Washington, D.C.: Center for Strategic and Budgetary Assessments, 2003), pp. 15–19.

39. Stephen Biddle, *Military Power: Explaining Victory and Defeat in Modern Battle* (Princeton, N.J.: Princeton University Press, 2004), pp. 28–51, 132–49, 190–208.

40. Ivo H. Daalder and Michael E. O'Hanlon, *Winning Ugly: NATO's War to Save Kosovo* (Washington, D.C.: Brookings, 2000).

41. Biddle, *Military Power*, pp. 55–60, 199–201.

42. Frederick W. Kagan, *Finding the Target: The Transformation of American Military Policy* (New York: Encounter Books, 2006), pp. 350–59.

43. Benjamin S. Lambeth, *Air Power Against Terror: America's Conduct of Operation Enduring Freedom* (Santa Monica, Calif.: RAND, 2005), p. 342.

44. See Ashton B. Carter, "Satellites and Anti-Satellites: The Limits of the Possible," *International Security*, vol. 10, no. 4 (Spring 1986), pp. 50–52; and David Wright, Laura Grego, and Lisbeth Gronlund, *The Physics of Space Security: A Reference Manual* (Cambridge, Mass.: American Academy of Arts and Sciences, 2005), pp. 40–46.

45. Barry D. Watts, *The Military Uses of Space: A Diagnostic Assessment* (Washington, D.C.: Center for Strategic and Budgetary Assessments, 2001), p. 123.

46. See, for example, Ed Kyle, "Space Launch Report, New Launchers: Space X Falcon," *Space Launch Report*, October 2006, available at www.geocities.com/launchreport/blog017.html [accessed January 10, 2008].

47. Tamar A. Mehuron, "2007 Space Almanac: The U.S. Military Space Operation in Facts and Figures," *Air Force Magazine* (August 2007), p. 82; Alan Collinson, "Briefing: Space Surveillance, Cutting the Clutter," *Jane's Defence Weekly*, January 16, 2008, p. 29; and Patterson Clark, "Current Missions," *The Washington Post*, September 25, 2008, p. G4. Much of the material in this section is taken from Michael E. O'Hanlon, *Neither Star Wars Nor Sanctuary: Constraining the Military Uses of Space* (Washington, D.C.: Brookings, 2004), p. 35.

48. Watts, *The Military Uses of Space*, p. 50.

49. Peter L. Hays, *United States Military Space: Into the Twenty-First Century* (Montgomery, Ala.: Air University Press, 2002), p. 133; and Joel R. Primack, "Debris and Future Space Activities," in James Clay Moltz, ed., *Future Security in Space: Commercial, Military, and Arms Control Trade-Offs*, Occasional Paper 10 (Monterey, Calif.: Monterey Institute of International Studies, 2002), pp. 18–20.

50. "Outlook/Specifications: Spacecraft," *Aviation Week and Space Technology*, January 28, 2008, p. 171.

51. "Outlook/Specifications: Spacecraft," *Aviation Week and Space Technology*, January 15, 2007, pp. 176–78, and January 28, 2008, pp. 170–72; Watts, *The Military Uses of Space*, pp. 42–43, 78; Craig Covault, "Secret NRO Recons Eye Iraqi Threats," *Aviation Week and Space Technology*, September 16, 2002, p. 23; Jeffrey T. Richelson, *America's Secret Eyes in Space: The U.S. Keyhole Spy Satellite*

Program (Harper and Row, 1990), pp. 130–32, 186–87, 206–8, 227, 236–38; O'Hanlon, *Neither Star Wars Nor Sanctuary*, pp. 42–53.

52. The ability of a satellite to image places on Earth not directly below it is limited by three factors: first, the ability of its camera or lens to swivel, second the need of the user for a certain minimum degree of resolution in the image (which often makes images taken at longer range less useful), and third the curvature of the Earth, which blocks distant regions from view. This last constraint is generally the most binding. To calculate the maximum range, for low-altitude satellites the formula is radar horizon = square root of (diameter of Earth × altitude of satellite). This follows directly from the Pythagorean theorem, drawing a right triangle with one side the radius of the Earth, a second side the distance from the satellite in question to the farthest point on Earth's surface within its view, and a third side from the center of the Earth to the satellite (this latter segment is the triangle's hypotenuse).

Using symbols, we can write more compactly $RH = \sqrt{(DA)}$. Since the diameter of the Earth is about 8,000 miles, a satellite at 200 miles' altitude can therefore "see" out about 1,250 miles (and an aircraft at just under eight miles' altitude can see about 250 miles).

53. "Outlook/Specifications: Spacecraft," *Aviation Week and Space Technology*, January 15, 2007, pp. 176–78, and January 28, 2008, pp. 170–72; Watts, *The Military Uses of Space*, pp. 42–43, 78; Craig Covault, "Secret NRO Recons Eye Iraqi Threats," *Aviation Week and Space Technology*, September 16, 2002, p. 23; Jeffrey T. Richelson, *America's Secret Eyes in Space: The U.S. Keyhole Spy Satellite Program* (Harper and Row, 1990), pp. 130–32, 186–87, 206–8, 227, 236–38; O'Hanlon, *Neither Star Wars Nor Sanctuary*, pp. 42–53.

54. "Outlook/Specifications: Spacecraft," *Aviation Week and Space Technology*, January 15, 2007, pp. 176–78, and January 28, 2008, pp. 170–72; and Mehuron, "2007 Space Almanac," pp. 87–89.

55. Mehuron, "2007 Space Almanac," p. 84.

56. Thomas A. Keaney and Eliot A. Cohen, *Gulf War Air Power Survey Summary Report* (Washington, D.C.: Government Printing Office, 1993), p. 193; Department of Defense, *Kosovo/Operation Allied Force After-Action Report* (Washington, D.C.: Department of Defense, 2000), p. 46; William B. Scott, "Milspace Comes of Age in Fighting Terror," *Aviation Week and Space Technology*, April 8, 2002, pp. 77–78; and Patrick Rayerman, "Exploiting Commercial SATCOM: A Better Way," *Parameters* (Winter 2003–2004), p. 55.

57. Jeremy Singer, "Laser Links in Space," *Air Force Magazine* (January 2008), p. 57.

58. Walt Faulconer, "Civilian Space Portfolio Assessment," briefing, Applied Physics Laboratory, Johns Hopkins University, Columbia, MD, April 22, 2008, p. 8.

59. Andrew E. Kramer, "Russia Challenges the U.S. Monopoly on Satellite Navigation," *The New York Times*, April 4, 2007, available at www.nytimes.com/2007/04/04/business/worldbusiness [accessed January 10, 2008].

60. Faulconer, "Civilian Space Portfolio Assessment," p. 8.

61. Kevin Pollpeter, *Building for the Future: China's Progress in Space Technology During the Tenth 5-Year Plan and the U.S. Response* (Carlisle, Pa.: Strategic Studies Institute, Army War College, March 2008) pp. 19–27; O'Hanlon, *Neither Star Wars Nor Sanctuary*, pp. 54–56; Steven A. Smith, "Chinese Space Superiority?: China's Military Space Capabilities and the Impact of Their Use in a Taiwan Conflict," Air War College, February 17, 2006, p. iii, available at www.au.af.mil/au/awc/awcgate/awc/smith.pdf [accessed January 10, 2008]; and Jeff Kueter, "China's Space Ambitions—And Ours," *The New Atlantis* (Spring 2007), pp. 7–8.

62. Smith, "Chinese Space Superiority?"; and Geoffrey Forden, "China's ASAT: No Space-age Pearl Harbor," *Wired* (January 11, 2008), available at http://blog.wired.com/defense/ [accessed January 16, 2008].

63. Joseph Post and Michael Bennett, *Alternatives for Military Space Radar* (Washington, D.C.: Congressional Budget Office, 2007), pp. ix–xxi.

64. Defense Science Board, *High Energy Laser Weapon Systems Applications* (Office of the Under Secretary of Defense for Acquisition, Technology, and Logistics, June 2001), pp. 49–54; and Elihu Zimet, "High-Energy Lasers: Technical, Operational, and Policy Issues," *Defense Horizons* 18 (Washington: National Defense University, Center for Technology and National Security Policy, October 2002), pp. 6–7 of 16, available at www.ndu.edu/inss/DefHor/DH18/DH_18.htm.

65. General Accounting Office, "Missile Defense: Knowledge-Based Process Would Benefit Airborne Laser Decision-Making," GAO-02-949T (July 16, 2002).

66. Missile Defense Agency Fact Sheet, "The Airborne Laser," Department of Defense, Washington, D.C., September 2007, available at www.mda.mil/mdalink/pdf/laser.pdf [accessed January 10, 2008].

67. Sandra I. Erwin, "Killing Missiles from Space: Can the U.S. Air Force Do It with Lasers?" *National Defense Magazine* (June 2001), pp. 3–5, available at www.nationaldefensemagazine.org/article.cfm?Id=513.

68. Celeste Johnson and Raymond Hall, *Estimated Costs and Technical Characteristics of Selected National Missile Defense Systems* (Washington, D.C.: Congressional Budget Office, 2002), pp. 20–27.

69. Steven M. Kosiak, *Arming the Heavens: A Preliminary Assessment of the Potential Cost and Cost-Effectiveness of Space-Based Weapons* (Washington, D.C.: Center for Strategic and Budgetary Assessments, 2007), pp. 38–40; and Bob Preston, Dana J. Johnson, Sean J. A. Edwards, Michael Miller, and Calvin Shipbaugh, *Space Weapons, Earth Wars* (Santa Monica, Calif.: RAND, 2002), pp. 40–49.

70. Jon Rosamond, "USN Admiral Says Satellite Kill Was 'One-Time Event,'" *Jane's Defence Weekly*, March 26, 2008, p. 8.

71. Bruce G. Blair, *Strategic Command and Control: Redefining the Nuclear Threat* (Washington, D.C.: Brookings, 1985), pp. 201–7; Ian Steer and Melanie Bright, "Blind, Deaf, and Dumb," *Jane's Defence Weekly*, October 23, 2002, pp. 21–23; Donald Rumsfeld, "Report of the Commission to Assess United States National Security Space Management and Organization" (Washington, D.C.: January 11, 2001), pp. 21–22; Carter, "Satellites and Anti-Satellites"; and Watts, *The Military Uses of Space*, p. 99.

72. Dennis Papadopoulos, "Satellite Threat Due to High Altitude Nuclear Detonations," briefing slides presented at Brookings on December 17, 2002, cited by permission from the author.

73. Philip E. Coyle, "Oversight of Ballistic Missile Defense (Part 3): Questions for the Missile Defense Agency," Testimony before the House Committee on Oversight and Government Reform, Subcommittee on National Security and Foreign Affairs, U.S. Congress, Washington, D.C., April 30, 2008, p. 12, available at www.cdi.org/pdfs/CoyleTestimonyApr08.pdf [accessed July 29, 2008].

74. Philip E. Coyle, "What Are the Prospects, What Are the Costs?: Oversight of Ballistic Missile Defense (Part 2)," Testimony before the House Committee on Oversight and Government Reform, Subcommittee on National Security and Foreign Affairs, U.S. Congress, Washington, D.C., April 16, 2008, p. 21, available at www.cdi.org/pdfs/CoyleHouseOversightGovtReform4_16_08.pdf [accessed July 29, 2008].

75. DoD News Briefing with Lt. Gen. Trey Obering, July 15, 2008, pp. 2–3, available at www.defenselink.mil/transcripts/transcript.aspx?transcriptid=4263 [accessed August 1, 2008]; and Ronald O'Rourke, "Sea-Based Ballistic Missile Defense—Background and Issues for Congress," *CRS Report for Congress* (Washington, D.C.: Congressional Research Service, May 23, 2008), pp. 13, 40, available at www.fas.org/sgp/crs/weapons/RL33745.pdf [accessed August 1, 2008].

76. David R. Tanks, *National Missile Defense: Policy Issues and Technological Capabilities* (Cambridge, Mass.: Institute for Foreign Policy Analysis, 2000), p. 3.3.

77. Curtis D. Cochran, Dennis M. Gorman, and Joseph D. Dumoulin, eds., *Space Handbook* (Maxwell Air Force Base, Alabama: Air University Press, 1985), pp. 3.27–30.

78. Thomas B. Cochran, William M. Arkin, and Milton M. Hoenig, *Nuclear Weapons Databook, Volume I: U.S. Nuclear Forces and Capabilities* (Ballinger Publishing, 1984), p. 107.

79. Tanks, *National Missile Defense*, p. 3.3.

80. See John Tirman, ed., *The Fallacy of Star Wars* (Vintage Books, 1984), pp. 52–65.

81. For more, see Stephen Weiner, "Systems and Technology," in Ashton B. Carter and David N. Schwartz, eds., *Ballistic Missile Defense* (Brookings, 1984), pp. 49–97; and Robert G. Nagler, *Ballistic Missile Proliferation: An Emerging Threat* (Arlington, Va.: System Planning Corporation, 1992), pp. 52–65.

82. For more, see David B. H. Denoon, *Ballistic Missile Defense in the Post–Cold War Era* (Westview Press, 1995), chaps. 3–5; and Department of Defense, "The Strategic Defense Initiative: Defense Technologies Study," reprinted in Steven E. Miller and Stephen Van Evera, eds., *The Star Wars Controversy* (Princeton University Press, 1986), pp. 291–322.

83. For more information, see the Federation of American Scientists' web site at (www.fas.org/spp/starwars/program [November 2000]).

84. See J. C. Toomay, *Radar Principles for the Non-Specialist* (Mendham, N.J.: SciTech Publishing, 1998), pp. 1–64.

85. David Mosher and Michael O'Hanlon, *The START Treaty and Beyond* (Congressional Budget Office, 1991), p. 148.

86. Mosher and O'Hanlon, *The START Treaty and Beyond*, pp. 167–71.

87. David Arthur and Robie Samanta Roy, *Alternatives for Boost-Phase Missile Defense* (Washington, D.C.: Congressional Budget Office, 2004), pp. 40–42; and RAND, *The Defense System Cost Performance Database: Cost Growth Analysis Using Selected Acquisition Reports* (Santa Monica, Calif.: RAND, 1996).

88. George N. Lewis and Theodore A. Postol, "Future Challenges to Ballistic Missile Defense," *IEEE Spectrum*, vol. 34 (September 1997), pp. 60–68.

89. The then–Martin-Marietta Corporation proposed fast-burn boosters back in the early 1980s; see Tirman, ed., *The Fallacy of Star Wars*, pp. 60–62.

90. See Andrew M. Sessler and others, *Countermeasures: A Technical Evaluation of the Operational Effectiveness of the Planned U.S. National Missile Defense System* (Cambridge, Mass.: Union of Concerned Scientists, April 2000), p. 42; and Gen. Larry Welch (ret.), chairman, and others, *Report of the Panel on Reducing Risk in Ballistic Missile Defense Flight Test Programs* (Department of Defense, February 27, 1998), p. 56 (www.fas.org/spp/starwars/program/welch/index.htm/ [November 2000]).

91. See the testimony of Richard L. Garwin and David C. Wright, "Ballistic Missiles: Threat and Response," Hearings before the Senate Committee on Foreign Relations, 106 Cong., 1 sess. (Government Printing Office, 2000), pp. 74–90.

92. Ann Scott Tyson, "U.S. Shoots Down Missile in Simulation of Long-Range Attack," *The Washington Post*, December 6, 2008, p. A2.

93. Ibid.

94. Bradley Graham, *Hit to Kill: The New Battle Over Shielding America from Missile Attack* (New York: Public Affairs, 2001), pp. 196–207.

95. David Wright, Laura Grego, and Lisbeth Gronlund, *The Physics of Space Security: A Reference Manual* (Cambridge, Mass.: American Academy of Arts and Sciences, 2005), pp. 98–100.

96. Michael Levi, *On Nuclear Terrorism* (Cambridge, Mass.: Harvard University Press, 2007), pp. 35–38, 52.

97. The same amount of deuterium undergoing fusion would produce about 25 kilotons of explosive force. See Samuel Glasstone, ed., *The Effects of Nuclear Weapons* (Washington, D.C.: U.S. Government Printing Office, 1962), pp. 5–6.

98. Edwin Lyman and Frank N. von Hippel, "Reprocessing Revisited: The International Dimensions of the Global Nuclear Energy Partnership," *Arms Control Today* (April 2008), p. 9.

99. Richard L. Garwin and Georges Charpak, *Megawatts and Megatons: A Turning Point in the Nuclear Age?* (New York: Alfred A. Knopf, 2001), pp. 58–61.

100. Levi, *On Nuclear Terrorism*, pp. 40–50.

101. Garwin and Charpak, *Megawatts and Megatons*, pp. 58–65.

102. Some of my arguments here first appeared in Michael O'Hanlon, "Resurrecting the Test-Ban Treaty," *Survival*, vol. 50, no. 1 (February–March 2008), pp. 119–32.

103. Zhang Hui, "Revisiting North Korea's Nuclear Test," *China Security*, vol. 3, no. 3 (Summer 2007), pp. 119–30.

104. Steve Fetter, *Toward a Comprehensive Test Ban* (Cambridge, Mass.: Ballinger, 1988), pp. 107–58.

105. America's Defense Monitor Interview with Richard Garwin, April 3, 1999, available at www.cdi.org/adm/1235/Garwin.html.

106. Jonathan Medalia, "The Reliable Replacement Warhead Program: Background and Current Developments," *CRS Report for Congress*, RL32929 (July 2007), pp. 4–9.

107. A. Fitzpatrick and I. Oelrich, "The Stockpile Stewardship Program: Fifteen Years On," *Federation of American Scientists* (April 2007), available at www.fas.org/2007/nuke/Stockpile_Stewardship_Paper.pdf [accessed January 9, 2008].

108. "At the Workbench: Interview with Bruce Goodwin of Lawrence Livermore Laboratories," *Bulletin of the Atomic Scientists* (July/August 2007), pp. 46–47.

109. National Nuclear Security Administration, "Reliable Replacement Warhead Program," March 2007, available at www.nnsa.doe.gov/docs/factsheets/2007/NA-07-FS-02.pdf.

110. Walter Pincus, "New Nuclear Warhead's Funding Eliminated," *The Washington Post*, May 24, 2007, p. A6.

111. John R. Harvey, "Nonproliferation's New Soldier," *Bulletin of the Atomic Scientists* (July/August 2007), pp. 32–33; and Jonathan Medalia, "The Reliable Replacement Warhead Program: Background and Current Develop-

ments," CRS Report for Congress RL32929, July 26, 2007, available at www.fas.org/sgp/crs/nuke/RL32929.pdf.

112. Michael A. Levi and Michael E. O'Hanlon, *The Future of Arms Control* (Washington, D.C.: Brookings, 2005), p. 28.

113. Michael A. Levi, "Dreaming of Clean Nukes," *Nature* 428, April 29, 2004, p. 892.

114. Charles L. Pritchard, *Failed Diplomacy: The Tragic Story of How North Korea Got the Bomb* (Washington, D.C.: Brookings, 2007), p. 203.

Key References and Suggestions for Further Reading

Arthur, David and Robie Samanta Roy, *Alternatives for Boost-Phase Missile Defense* (Washington, D.C.: Congressional Budget Office, 2004).

Biddle, Stephen, *Military Power: Explaining Victory and Defeat in Modern Battle* (Princeton, N.J.: Princeton University Press, 2004).

Boot, Max, *War Made New: Technology, Warfare, and the Course of History, 1500 to Today* (New York: Gotham Books, 2006).

Carter, Ashton B., "Satellites and Anti-Satellites: The Limits of the Possible," *International Security*, vol. 10, no. 4 (Spring 1986).

Diamond, Jared, *Guns, Germs, and Steel: The Fates of Human Societies* (W.W. Norton, 1997).

Fetter, Steve, *Toward a Comprehensive Test Ban* (Cambridge, Mass.: Ballinger, 1988).

Freedman, Lawrence, *The Revolution in Strategic Affairs*, Adelphi Paper 318, International Institute for Strategic Studies (Oxford, England: Oxford University Press, 1998).

Garwin, Richard L. and Georges Charpak, *Megawatts and Megatons: A Turning Point in the Nuclear Age?* (New York: Alfred A. Knopf, 2001).

Glasstone, Samuel, ed., *The Effects of Nuclear Weapons* (Washington, D.C.: U.S. Government Printing Office, 1962).

Graham, Bradley, *Hit to Kill: The New Battle Over Shielding America from Missile Attack* (New York: Public Affairs, 2001).

Johnson, Celeste, and Raymond Hall, *Estimated Costs and Technical Characteristics of Selected National Missile Defense Systems* (Washington, D.C.: Congressional Budget Office, 2002).

Johnson, Stuart E. and Martin C. Libicki, eds., *Dominant Battlespace Knowledge* (Washington, D.C.: National Defense University, 1996).

Kagan, Frederick W., *Finding the Target: The Transformation of American Military Power* (New York: Encounter Books, 2006).

Kosiak, Steven M., *Arming the Heavens: A Preliminary Assessment of the Potential Cost and Cost-Effectiveness of Space-Based Weapons* (Washington, D.C.: Center for Strategic and Budgetary Assessments, 2007).

Krepinevich, Andrew, Jr., "Cavalry to Computer: The Pattern of Military Revolutions," *National Interest*, no. 37 (Fall 1994), pp. 31–36.

Krepinevich, Andrew, Jr., Barry Watts, and Robert Work, *Meeting the Anti-Access and Area-Denial Challenge* (Washington, D.C.: Center for Strategic and Budgetary Assessments, 2003).

Levi, Michael, *On Nuclear Terrorism* (Cambridge, Mass.: Harvard University Press, 2007).

Lyman, Edwin, and Frank N. von Hippel, "Reprocessing Revisited: The International Dimensions of the Global Nuclear Energy Partnership," *Arms Control Today* (April 2008).

Murray, Williamson, "Thinking About Revolutions in Military Affairs," *Joint Forces Quarterly* (Summer 1997), pp. 69–76.

National Defense Panel, *Transforming Defense: National Security in the 21st Century* (Washington, D.C.: National Defense Panel, 1997).

Preston, Bob, Dana J. Johnson, Sean J. A. Edwards, Michael Miller, and Calvin Shipbaugh, *Space Weapons, Earth Wars* (Santa Monica, Calif.: RAND, 2002).

Sessler, Andrew M. and others, *Countermeasures: A Technical Evaluation of the Operational Effectiveness of the Planned U.S. National Missile Defense System* (Cambridge, Mass.: Union of Concerned Scientists, April 2000).

Stillion, John and David T. Orletsky, *Airbase Vulnerability to Conventional Cruise-Missile and Ballistic-Missile Attacks: Technology, Scenarios, and U.S. Air Force Responses* (Santa Monica, Calif.: RAND, 1999).

Van Creveld, Martin, *Technology and War: From 2000 B.C. to the Present* (New York: Free Press, 1991).

Vickers, Michael G. and Robert C. Martinage, *The Revolution in War* (Washington, D.C.: Center for Strategic and Budgetary Assessments, 2004).

von Hippel, Frank, *Citizen Scientist* (New York: Touchstone, 1991).

Watts, Barry D., *The Military Uses of Space: A Diagnostic Assessment* (Washington, D.C.: Center for Strategic and Budgetary Assessments, 2001).

Welch, Gen. Larry (ret.), and others, *Report of the Panel on Reducing Risk in Ballistic Missile Defense Flight Test Programs* (Department of Defense, February 27, 1998), p. 56 (www.fas.org/spp/starwars/program/welch/index.htm/ [November 2000]).

Wright, David, Laura Grego, and Lisbeth Gronlund, *The Physics of Space Security: A Reference Manual* (Cambridge, Mass.: American Academy of Arts and Sciences, 2005).

CONCLUSION

As an effort to collect, assess, and present simplified tools of defense analysis, this book does not require a major conclusion to draw an overall argument together. There is no single core policy purpose for the book, except to help inform a debate on critically important matters of war and peace, of strategies for combat and for deterrence, and of national resource allocation.

It is useful, however, to repeat some of the central observations and conclusions of the book that do bear on contemporary policy questions fairly directly. They are not policy recommendations themselves, but are sufficiently well established by analysis to be relevant to practical debates. The following does not comprise an exhaustive list, but is an attempt to capture the spirit and approach of the text. As with the main organization of the book, the topics fall into four areas: budget analysis, combat modeling, logistics, and defense technology.

- Despite the amount of effort and money devoted to the undertaking, defense budget analysis remains imprecise. This is partly because the course of possible future wars cannot easily be predicted, partly because the expense involved in inventing a new technology cannot be accurately forecast before the fact. It is also because of sloppiness (or political motivation) in how calculations are sometimes presented. A goal of this book is to help reduce sloppiness and expose political or parochial motivations when such calculations are done and discussed.
- To avoid tendentious defense budgeting, it is always important to understand the budgetary details of a defense policy proposal. Are the projected costs or savings for a single year or for the long term? Are sunk costs being included in discussions of a future budget option when they should not be? Are procurement costs

for a system not yet even developed being viewed as firm and reliable when they should not be?

- Some basic budgetary rules of thumb are worth bearing in mind—weapons generally cost about 20 to 50 percent more to produce than predicted before they are developed or procured; American troops cost about $100,000 a year each in personnel costs alone (and a roughly comparable amount to support and train); deploying troops abroad costs extra, above and beyond the normal Pentagon budget (which normally does not include funds for wars or other major operations abroad), to the tune of more than $500,000 per Marine or soldier per year in Iraq and Afghanistan. But these rules are rough, and are not necessarily good indicators of future realities. For example, the last figure is perhaps twice what might have been predicted based on previous operations (the U.S. military operation in Bosnia was forecast to cost about $100,000 per soldier per year in 1995–1996 and wound up costing twice that itself).

- Much of the political rhetoric surrounding the state of the current American military is overblown. Claims that the wars have broken the force cannot be substantiated by the data. Equipment stocks on the whole remain in generally reasonable condition (even if 20 to 25 percent of ground combat equipment is deployed and therefore unavailable for other missions at any given time, with much higher percentages of specific assets such as up-armored HMMWVs committed in Iraq and Afghanistan). In addition, the quality of American military personnel remains very high by historical standards, even taking into account all the recent challenges of recruiting and retention.

- Those dismissive of the strains on the force go too far." Fairness considerations alone suggest serious questioning about whether the nation is asking too much of its men and women in uniform. Even if the overall state of the military remains good—in terms of the aptitude levels of new recruits, or the experience and retention levels of typical soldiers and Marines—current trends are quite worrisome. Suicides and divorces are substantially higher than in the recent past, if not necessarily beyond overall historical (or societal) norms, as are waivers for former felons joining the Army and Marine Corps. Plus, the sheer strain on people who have been deployed two and three and four times represents a potential ticking time bomb. There is no readiness crisis today,

and no clear trend towards one, but given the strains on the force, this conclusion is always subject to change.
- American military spending dominates world figures. However, the Iraqi resistance has shown what a group with less than 1 percent of the total revenue of the U.S. military can do on the battlefield, under certain circumstances, against the world's only superpower. This suggests that defense budgets are hardly the main determinants of battle outcomes. On the other hand, when the United States increased its resource allocation to the Iraq War by some 25 percent in 2007, it began to achieve synergies of effort, and an overall effectiveness, that had eluded it before. Of course, its strategy changed as well. But the sheer application of additional resources was an important aspect of that strategy. This suggests that, under certain circumstances, it may matter a great deal whether the United States outspends its foe by a factor of 10, or 20, or 100 in a given conflict. In the end, budget comparisons are poor predictors of combat outcomes.
- Combat models are better predictors of outcomes than are budget comparisons. These are fraught with risks, however, and imprecisions, too. Several questions must be kept in mind when using them. Do they employ the wrong historical battle or war as an analogy for a future war? Do they involve such detailed computer simulations as to appear more scientific than they really are—and obscure the key assumptions that go into them about how long adversaries will fight and the tactics they will employ?
- A good and simple rule of thumb to avoid oversimplifying predictions about future combat is to be sure a range of plausible combat outcomes is always presented. Failure to do so usually indicates overconfidence about how well war, an inherently human and unpredictable venture, can be accurately forecast.
- Generally speaking, if a combat model predicts the outcome of a war within a factor of two to five—in terms of the duration of fighting, the relative casualties of the two sides, and the overall amount of devastation and destruction—it will have done reasonably well. In other words, making a prediction that is wrong by "only 100 percent" (for example, forecasting losses to one's own forces of 100 soldiers but having 200 killed in the actual operation) is actually rather precise by the standards of the field. In fact, being off by 300 or 400 percent is not bad—provided that

the basic flow of battle, and the winner and loser, are not mistakenly predicted.
- It is worth bearing in mind that historically, outnumbered attackers have tended to win around half the wars they have initiated. So simple quantitative rules of thumb, such as the idea that an attacker should have three times the battle strength of a defender to prevail, are misleading at best. They may sometimes be useful guides for force planners, but are rarely if ever solid bases for prognostication.
- Military models, despite these limitations, have great uses. They tended to predict rapid U.S. victories in places like Panama and in Operation Desert Storm. They raised warnings about resultant conflicts in the Kosovo campaign and the second Iraq war. That was less because models yielded accurate predictions in these cases, and more because it was difficult to see how to apply standard models to these scenarios. That fact should have sent up red flags for anyone hazarding a prediction of either war's outcome with great confidence.
- In Iraq and Afghanistan, quantitative rules of thumb and models have consistently highlighted the uncomfortable fact that coalition force levels have been very small by historical standards. Even if those historical standards did not provide an exact answer about how many forces we actually needed in Iraq and Afghanistan, they certainly should have made careful analysts aware of the dangers of the operations as planned.
- Models are useful in other situations, too. They can, for example, help sketch out the hypothetical vulnerability of a city like Seoul to artillery barrage from North Korea. Models can reveal the difficulties a country such as Israel is likely to have in wars like that in Lebanon in 2006 (again, since we know how hard it is to find and target enemy mortars and artillery set against complex terrain). Indeed, in its late 2008 attacks on Hamas in the Gaza Strip, Israel seemed to emphasize attacks on Hamas leadership at least as much as on weaponry—and to have made greater efforts to achieve surprise in its initial strikes so as to maximize effectiveness. Or for a possible war between Taiwan and China (which could also involve the United States), models can make it clear why the PRC would have great trouble in executing a successful all-out amphibious assault—while also demonstrating that China might have numerous other military options short of all-out war.

- The old adage that while civilians think strategy, generals think logistics, remains as true as ever. Modern military forces in large operations typically require a ton of supplies per day for every ten soldiers (plus or minus a factor of two to three). For large ground operations, at least two-thirds of the weight is typically in the form of fuel and water, but there are large requirements for ammunition, spare parts, food, and other solid supplies, too (not to mention the weight of the weapons and vehicles themselves—a U.S. heavy division of about 18,000 troops weighs over 100,000 tons). Sustaining such a logistics flow is beyond the capacity of almost all militaries in the world save those of the United States and, to a much lesser degree, Britain and France. Other countries typically cannot operate far from home territory unless they contract for supplies and the transportation of those supplies, or unless they have many months to prepare a given mission. Even then, their capacities are typically limited to a few hundred troops, or at most a few thousand.
- For all the recent talk of moving to lighter and more maneuverable forces, that are less attached to their logistics tails than militaries of the past, actual trends have continued to go in the other direction, towards heavier forces needing more supplies. In 2002, a different trend seemed underway, since the Kosovo and Afghanistan wars both accomplished some of their goals with modest American capabilities. But the Iraq War has trumped these earlier experiences and reminded everyone how so much about combat has *not* changed.
- This conclusion about logistics underscores a broader reality regarding military technology. On the one hand it is very impressive, and always changing. On the other, it does not allow truly radical and rapid shifts in how war is fought very often. The hypothesis that a modern revolution in military affairs was occurring, or at least was within reach, may have helped American strategists avoid complacency in the early years after the Cold War. However, aspects of that debate also arguably contributed to the U.S. sloppiness in preparing for the invasion of Iraq.

Military analysis is not an exact science. To return to the wisdom of Sun Tzu, and paraphrase the great Chinese political philosopher, it is at least as close to art. But many logical methods offer insight into military problems—even if solutions to those problems ultimately require the use of judgment and of broader political and strategic considerations as well. Military affairs may not be as amenable to quantification and formal methodological treatment as economics, for example. However, even if our main goal in analysis is generally to illuminate choices, bound problems, and rule out bad options—rather than arrive unambiguously at clear policy choices—the discipline of military analysis has a great deal to offer. Moreover, simple back-of-the envelope methodologies often provide substantial insight without requiring the churning of giant computer models or access to the classified data of official Pentagon studies, allowing generalists and outsiders to play important roles in defense analytical debates.

We have seen all too often (in the broad course of history as well as in modern times) what happens when we make key defense policy decisions based solely on instinct, ideology, and impression. To avoid cavalier, careless, and agenda-driven decision-making, we therefore need to study the science of war as well—even as we also remember the cautions of Clausewitz and avoid hubris in our predictions about how any war or other major military endeavor will ultimately unfold.

APPENDIX

Figures and Tables

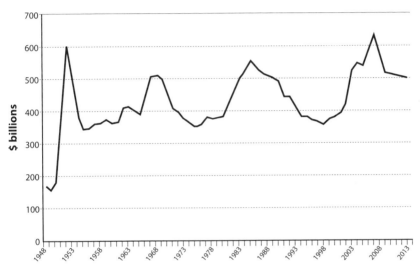

FIGURE A.1: Department of Defense Annual Budget Authority, 1948–2013 (Constant FY 2009 $)
Source: DoD, "National Defense Budget Estimates for FY 2009," pp. 109–14.

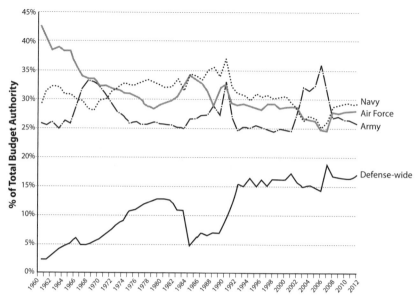

FIGURE A.2: Department of Defense, Budget Authority by Service, FY 1960–2013
Source: http://www.defenselink.mil/comptroller/defbudget/fy2009/FY09Greenbook/greenbook_2009_updated.pdf, (pp. 67–72).

Note: For 1991, the official percentage of the budget authority for "Defense-wide" spending was −3 percent. This was due to foreign contributions to Operation Desert Storm. This is not depicted in the graph above since the curve was flattened between the years 1990 and 1992.

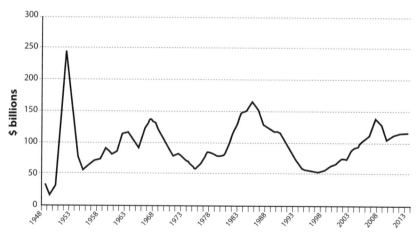

FIGURE A.3: Department of Defense Annual Budget Authority for Procurement, FY 1948–2013 (Constant FY 2009 $)
Source: U.S. Department of Defense, "National Defense Budget Estimates for FY 2009," pp. 103–8.

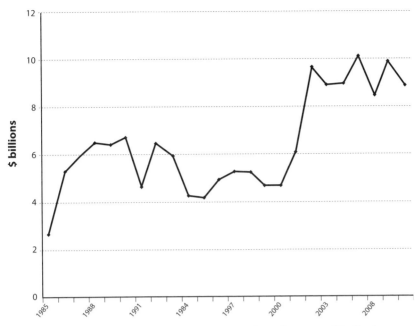

FIGURE A.4: Historical Funding for the Missile Defense Agency and Its Predecessors (Constant FY 2009 $)

Source: www.mda.mil/mdaLink/pdf/histfunds.pdf.

Note: Historical funding levels are for Strategic Defense Initiative Organization (SDIO), the Ballistic Missile Defense Organization (BMDO), and the Missile Defense Agency (MDA). SDIO was the predecessor to the MDA. Some missle defense funding is outside these agencies.

TABLE A.1
U.S. Military Annual Active Duty Personnel End Strength, 1960–2009

YEAR	TOTAL	Army	Navy	Marines	Air Force
1960	2,492,037	877,749	624,895	175,919	813,474
1961	2,552,912	893,323	641,995	185,165	832,429
1962	2,687,690	962,712	662,837	192,049	870,092
1963	2,695,240	961,211	668,626	189,937	875,466
1964	2,690,141	972,546	670,160	189,634	857,801

(continued)

TABLE A.1 (*cont.*)

YEAR	TOTAL	Army	Navy	Marines	Air Force
1965	2,723,800	1,002,427	690,162	198,328	832,883
1966	3,229,209	1,310,144	740,646	280,641	897,778
1967	3,411,931	1,468,754	749,299	299,501	894,377
1968	3,489,588	1,516,973	759,163	308,138	905,314
1969	3,449,271	1,514,223	764,867	311,627	858,554
1970	2,983,868	1,293,276	677,152	246,153	767,287
1971	2,626,785	1,050,425	615,767	204,738	755,855
1972	2,356,301	849,824	593,135	199,624	713,718
1973	2,231,908	791,460	566,653	192,064	681,731
1974	2,157,023	784,128	546,464	192,174	634,257
1975	2,104,795	775,301	532,270	195,683	601,541
1976	2,083,581	782,668	527,781	189,851	583,281
1977	2,074,543	782,246	529,895	191,707	570,695
1978	2,062,404	771,624	530,253	190,815	569,712
1979	2,027,494	758,852	523,937	185,250	559,455
1980	2,050,826	777,036	527,352	188,469	557,969
1981	2,082,897	781,473	540,502	190,620	570,302
1982	2,108,612	780,391	552,996	192,380	582,845
1983	2,123,349	779,643	557,573	194,089	592,044
1984	2,138,157	780,180	564,638	196,214	597,125
1985	2,151,032	780,787	570,705	198,025	601,515

(*continued*)

TABLE A.1 (cont.)

YEAR	TOTAL	Army	Navy	Marines	Air Force
1986	2,169,112	780,980	581,119	198,814	608,199
1987	2,174,217	780,815	586,842	199,525	607,035
1988	2,138,213	771,847	592,570	197,350	576,446
1989	2,130,229	769,741	592,652	196,956	570,880
1990	2,046,144	732,403	581,856	196,652	535,233
1991	1,986,259	710,821	570,966	194,040	510,432
1992	1,807,177	610,450	541,883	184,529	470,315
1993	1,705,103	572,423	509,950	178,379	444,351
1994	1,610,490	541,343	468,662	174,158	426,327
1995	1,518,224	508,559	434,617	174,639	400,409
1996	1,471,722	491,103	416,735	174,883	389,001
1997	1,438,562	491,707	395,564	173,908	377,385
1998	1,406,830	483,880	382,338	173,142	367,470
1999	1,385,703	479,426	373,046	172,641	360,590
2000	1,384,338	482,170	373,193	173,321	355,654
2001	1,385,116	480,801	377,810	172,934	353,571
2002	1,411,634	486,542	383,108	173,733	368,251
2003	1,434,377	499,301	382,235	177,779	375,062
2004	1,426,836	499,543	373,197	177,480	376,616
2005	1,389,394	492,728	362,941	180,029	353,696
2006	1,384,968	505,402	350,197	180,416	348,953

(continued)

Table a.1 (cont.)

YEAR	TOTAL	Army	Navy	Marines	Air Force
2007	1,379,551	522,017	337,547	186,492	333,495
2008*	1,385,122	531,526	331,785	193,040	328,771

Source: Department of Defense, Military Personnel Statistics web site

Note: Unless otherwise noted, figures are as of September 30 for each year, which corresponds to the end of the fiscal year, and do not include activated reservists.

*As of June 30, 2008

Table a.2

U.S. Troops Based in Foreign Countries (As of June 30, 2008)

Country or Region	Total
Europe	
Belgium	1,301
Germany	55,145
Italy	9,515
Netherlands	552
Portugal	792
Serbia (includes Kosovo)	1,289
Spain	1,238
Turkey	1,570
United Kingdom	9,613
Afloat	848
Other	958
TOTAL	82,821
Former Soviet Union	144

(*continued*)

TABLE A.2 (cont.)

Country or Region	Total
East Asia and Pacific	
Japan	32,966
Korea	25,374
Afloat	8,975
Other	774
TOTAL	68,089
North Africa, Near East, and South Asia	
Bahrain	1,504
Afloat	2,222
Other	1,648
TOTAL	5,374
Sub-Saharan Africa	
Djibouti	1,900
Other	325
TOTAL	2,225
Western Hemisphere	
Cuba (Guantanamo)	980
Other	1,114
TOTAL	2,094
Undistributed	
Ashore	112,910
Afloat	6,168

(continued)

TABLE A.2 (cont.)

Country or Region	Total
TOTAL	119,078
TOTAL- Foreign Countries	279,825
Ashore	261,602
Afloat	18,223
DEPLOYMENTS	
Operation Iraqi Freedom TOTAL	183,100
Operation Enduring Freedom	31,700
TOTAL	214,800
Deployed from Locations for OIF/OEF (outside U.S.)	
Germany	18,300
Italy	2,200
Japan	2,400
Korea	100
United Kingdom	950
TOTAL	23,950

Source: DoD Personnel Statistics; http://siadapp.dmdc.osd.mil/personnel/MILITARY/history/hst0806.pdf.

Note: Only countries with at least 500 troops are listed individually.

TABLE A.3
Selected Acquisition Report (SAR) Program Acquisition Cost Summary
(As of December 31, 2007 in Millions of Dollars)

Weapons System	Base Year	Current Estimate Base Year $	Current $	Quantity
Army				
Apache Block III (AB3)	2006	7,158	8,996	639
ARH	2005	5,260	6,337	512
ATIRCM/CMWS	2003	4,170	4,816	3,589
Black Hawk upgrade	2005	18,935	24,043	1,235
Bradley upgrade	2001	8,570	9,695	2,568
CH-47F	2005	11,516	13,350	513
Excalibur	2007	2,233	2,465	30,388
FBCB2	2005	3,220	3,371	73,463
FCS	2003	112,425	159,320	15
FMTV	1996	16,517	20,676	83,185
GMLRS	2003	4,718	6,008	43,795
HIMARS	2003	1,797	2,049	381
Javelin	1997	4,694	4,924	25,463
JLENS	2005	6,089	7,500	16
Longbow Apache	1996	9,826	11,183	671
AUH	2006	1,820	2,090	345
Patriot PAC-3	2002	8,387	8,525	969
Patriot/MEADS CAP - Fire Unit	2004	15,808	21,780	48

(*continued*)

TABLE A.3 (cont.)

Weapons System	Base Year	Current Estimate Base Year $	Current $	Quantity
Patriot/ MEADS CAP - Missile	2004	6,027	8,116	1,528
Stryker	2004	14,255	15,691	3,537
WIN-T increment 1	2007	3,798	3,860	1,677
WIN-T increment 2	2007	3,446	3,871	1,893
Subtotal, Army		270,667	348,667	
Navy				
ADS (AN/WQR-3)	2005	552	529	—
AGM-88E AARGM	2003	1,426	1,710	1,911
AIM-9X	1997	2,664	3,396	10,142
CEC	2002	4,207	4,531	306
CH-53K	2006	15,025	18,708	156
Cobra Judy replacement	2003	1,453	1,630	1
CVN 21	2000	24,987	35,119	3
CVN 68	1995	5,279	6,259	1
DDG 1000	2005	25,090	28,887	7
DDG 51	1987	46,418	62,756	62
E-2D AHE	2002	13,394	17,431	75
EA-18G	2004	7,578	8,649	85
EFV	2007	13,164	15,860	593
ERM	2005	1,288	1,521	15,100
F/A-18 E/F	2000	43,258	46,345	493

(continued)

TABLE A.3 (cont.)

Weapons System	Base Year	Current Estimate Base Year $	Current $	Quantity
H-1 upgrades (4BW/4BN)	1996	6,750	8,728	284
JSOW - Baseline/BLU-108	1990	1,476	1,862	3,334
JSOW - Unitary	1990	1,777	2,725	7,000
LCS	2004	2,595	2,849	2
LHA replacement	2006	3,079	3,368	1
LPD 17	1996	11,508	14,242	9
MH-60R	2006	11,279	12,139	254
MH-60S	1998	6,504	7,843	271
MUOS	2004	5,667	6,682	6
NMT	2002	1,677	2,103	305
P-8A (MMA)	2004	26,183	32,853	113
RMS	2006	1,380	1,550	108
SM-6	2004	4,693	5,954	1,200
SSDS	2004	557	669	42
SSGN	2002	3,867	4,109	4
SSN 774 (Virginia Class)	1995	63,752	91,965	30
T-45TS	1995	6,735	6,828	223
Tactical Tomahawk	1999	3,706	4,375	3,292
T-AKE	2000	4,618	5,715	12
Trident II missile	1983	26,382	38,817	561
V-22	2005	50,473	54,227	458

(continued)

TABLE A.3 (cont.)

Weapons System	Base Year	Current Estimate Base Year $	Current $	Quantity
VH-71	2003	5,732	6,750	28
VTUAV	2006	1,875	2,158	177
Subtotal, Navy		458,047	571,841	
Air Force				
AEHF	2002	6,738	7,362	4
AMRAAM	1992	13,156	14,881	13,953
B-2 EHF increment I	2007	636	681	21
B-2 RMP	2004	1,094	1,225	21
C-130 AMP	2000	4,521	5,800	222
C-130J	1996	9,805	12,029	134
C-17A	1996	58,665	62,307	190
C-5 AMP	2006	1,377	1,405	112
C-5 RERP	2000	8,478	11,131	111
F-22	2005	66,992	64,540	184
FAB-T	2002	2,963	3,622	222
GBS	1997	727	806	1,121
Global Hawk (RQ-4A/B)	2000	8,102	9,741	54
JASSM	1995	4,466	6,066	5,006
JDAM	1995	4,522	5,260	201,993
JPATS	2002	4,915	5,534	768
Minuteman III GRP	1993	2,095	2,428	652

(continued)

TABLE A.3 (cont.)

Weapons System	Base Year	Current Estimate Base Year $	Current $	Quantity
Minuteman III PRP	1994	2,190	2,602	601
MP RTIP	2000	1,115	1,225	–
MPS	2004	1,394	1,583	1
NAS	2005	1,424	1,491	91
NAVSTAR GPS - Space and Control	2000	5,963	6,306	33
NAVSTAR GPS - User Equipment	2000	1,791	2,094	–
NPOESS	2002	9,363	11,140	4
SBIRS High	1995	9,559	11,556	4
SDB I	2001	1,252	1,477	24,070
WGS	2001	1,764	1,951	5
Subtotal, Air Force		235,065	256,240	
DoD				
BMDS	2002	89,398	102,912	–
CHEM DEMIL- ACWA	1994	5,499	7,992	3,136
CHEM DEMIL- CMA	1994	22,459	27,423	29,060
DIMHRS	2007	850	819	1
F-35 (JSF)	2002	210,015	298,843	2,456
JTRS GMR	2002	14,243	20,536	86,652
JTRS HMS	2004	2,672	3,367	95,961
JTRS NED	2002	1,743	1,962	–

(continued)

Table A.3 (cont.)

Weapons System	Base Year	Current Estimate Base Year $	Current $	Quantity
MIDS	2003	2,289	2,373	3,807
Subtotal, DoD		349,168	466,226	
GRAND TOTAL		1,312,947	1,642,974	

Source: http://www.acq.osd.mil/ara/am/sar/2007-DEC-SARSUMTAB.pdf.

Note: Totals may not add up exactly as shown due to rounding. Each weapon is assigned a base year based on key milestones in its development; costs as expressed in "base year dollars" are measured in that base year's constant dollars. Procurement costs, as well as research, development, test, and evaluation costs, are included.

INDEX

Afghanistan wars, 41, 146, 185–86, 230–31, 244–47
Africa Command (AFRICOM), 6, 156
Airborne Laser (ABL), 194–96, 207–9
air combat, 88–89, 129–30, 158–59
Air Combat Command, 6
Air Force War College, 3
Air Mobility Command, 3
air superiority, 131
Al Qaeda, 112–13, 180, 185
Albania, 34
Algeria, 110
amphibious assault, 85–96
Anti-Ballistic Missile (ABM) Treaty, 201
anti-satellite (ASAT) weapons, 195–99
antisubmarine warfare (ASW), 100–103, 122–24
Anzio, Italy, 87
Apache helicopter, 34
Applied Physics Laboratory, Johns Hopkins University, 3
Arab-Israeli War, 1967, 65, 158
Armed Forces Qualification Test (AFQT), 37
armored division equivalent (ADE), 73
Army Concepts Analysis Agency, 64
Army force structure, 127–28
Army War College, 3
Art of War, The, 1
artillery barrages, 97

B-2 bomber, 47–49
Base Force, Bush Administration, 21
bases, military, 152–62, 254–56; in Africa, 156; in Asia, 154–55, 161–62; in Europe, 152–54, 160–61; in Middle East, 155–56
Bay of Pigs, 87
benchmarks in counterinsurgency, stabilization missions, 110–14
Betts, Richard, 33
Blainey, Geoffrey, 63

blockades, naval, 96–103
Bosnia, 17, 97
budget authority, 11
budget breakdowns, Department of Defense, 13–15, 18–27, 28–29, 249–51, 257–62
budget comparisons, international, 43–47
Builder, Carl, 5

C Ratings for readiness, 35
C-5 aircraft, 148–49
C-17 aircraft, 148–49
C-130 aircraft, 149–50
C-141 aircraft, 148–49
CH-53 helicopter, 149
Campbell, Jason, 114
Cape Canaveral, Florida, 191
Carafano, Jim, 182
Central Command (CENTCOM), 6, 155–56
Central Intelligence Agency (CIA), 7, 186
chemical weapons, 95
China Lake Naval Air Weapons Station, 3
China-Taiwan conflict, 88–103, 246
Chinese military spending, 43–47
Chinese missiles, 88–89, 98–100, 211–12
Chinese satellites, 192–93, 198–99
Clausewitz, Carl von, 1, 248
Clinton administration, 34
Cold War military spending, 8–9
combat exchange ratio, 77
Combined Action Program (CAP), U.S. Marine Corps in Vietnam, 111
command and control, nuclear, 105
communications technologies, 179–80, 186, 192
Comprehensive Test Ban Treaty (CTBT), 218–26, 228–29
Congo, Democratic Republic of, 116
Congressional Budget Office (CBO), 17, 18, 49, 151
contractors, Defense, 18–19
convoy escort, naval, 100–103

cost growth, 28–30
countermeasures for missile defense, 210–14

Dahlgren Surface Warfare Facility, 3
Davis, Colonel M. Thomas, 12
Defense Science Board, 170, 179
Department of Energy nuclear weapons budget and activities, 12, 222
Diplomatic (State Department) Budget, 9
discretionary budgets, 11
Dunnigan, James, 4
Dupuy, Trevor, 73, 80–85, 121–22

engines, 180–83
Epstein, Joshua, 64, 66, 72–80, 117–19, 121, 146
European Command (EUCOM), 6, 152–54

Falklands (Malvinas) War, 87, 93–94
FBCB2 (Force 21 Battle Command Brigade and Below), 179
Foreign Assistance Budget, 9, 10
fratricide, nuclear, 105
Future Years Defense Program (FYDP), 11, 12

Gallipoli, Turkey, 87
Garwin, Richard, 221
G.E.D. degrees, 36–37
Global Positioning System (GPS) Satellites, 48, 179, 183–85, 189–93
Goldwater-Nichols Act of 1986
Guam, 49–51, 154–57
Gudgel, Andrew, 182

Hale, Robert, 18
Hamas, 246
High Mobility Multipurpose Wheeled Vehicles (HMMWVs), 42, 244
Hiroshima bomb, 216–19, 223, 227
history of military innovation and revolution, 174–76
Homeland Security budget, 9, 10
homeporting of ships, 159
How to Make War, Dunnigan's, 4
Hughes, Captain Wayne, Jr., 92–93, 102

Inchon, Korea, 87
India, 114–18
Indonesia, 116
Institute for Defense Analyses, 3, 66
intercontinental ballistic missiles (ICBMs), 103–7, 205–6

International Institute for Strategic Studies, 7, 8, 43
Iran hostage rescue attempt, 33
Iran-Iraq war, 95
Iran, nuclear weapons, 225
Iran Strait of Hormuz scenario, 122–24
Iraq War, Operation Iraqi Freedom, 34, 41, 65, 83–85, 112–14, 142, 185–86, 226–27, 230–31, 244–47
Israel, 246

JASON group, 170
Joint Chiefs of Staff (JCS), 6
Joint Direct Attack Munition (JDAM), 48, 80, 183–85
Joint Forces Command, 6
Joint Improvised Explosive Device Defeat Organization (JIEDDO), 6
Joint Staff, 3
Joint Surveillance and Target Attack Radar System (JSTARS), 76, 120

KC-10 aircraft, 148–49
Kashmir, 107, 114–16
Kaufmann, William, 10, 18–27, 51, 53
Kirtland Air Force Base, 3
Korean conflict, 97
Kosovo War, NATO, 34, 74, 171, 183, 185, 246–47
Kugler, Richard, 72–80, 117–19, 121
Kugler-Posen "FEBA Expansion" Model, 72–80, 117–19, 121
Kuperman, Alan, 151

Labs, Eric, 49
Lanchester equations, 67–71
large medium-speed roll-on/roll-off (LMSR) ships, 150
laser technology, 194–97
Lebanon, 33, 185
Libya, 34
Lincoln Labs, MIT, 170
Livermore National Laboratory, 3
logistics, military, 141–52
Los Alamos National Laboratory, 3
low-cost autonomous attack system (LOCAAS), 183
low-density high-demand units, 38

major procurement, 30
mandatory budgets, 11
Masks of War, Builder's, 5

Massive Ordnance Penetrator, 184
Maverick munition, 78
McNamara Programs, defense budget breakdown, 13, 18
metrics in counterinsurgency, stabilization missions, 110–14
microsatellites, 198–99
midcourse missile defense system, 201–2, 207–9, 213
Military Balance, The, 7
military force comparisons, international, 45
military personnel, U.S., over time, 251–54
military readiness, 31–43, 52–53
Military Sealift Command, 3
military technology, 177–87
Mine Resistant Ambush Protected (MRAP) vehicles, 17, 42, 148
mine warfare, 85–96, 122–24
minor procurement, 30
missile defense, 123–24, 201–15
Missile Defense Agency, 202, 214
missile flight trajectories, 202–6
multiple independently-targetable reentry vehicles (MIRVs), 204, 210–12, 214

Nagasaki bomb, 219, 227
National Defense Panel, 180–81
National Defense Strategy (of Secretary of Defense), 4
National Military Strategy (of Joint Chiefs), 12
National Security Act of 1947, 5
National Security Agency, 7
National Security Budget, 10
National Security Strategy, 12
Naval War College, 3
Navy Theater Wide (NTW) defense system, 201–2, 207–9
Network-Centric Warfare, 179
Nigeria, 116
no-fly zone, Iraqi, 123
Normandy, 87
North Korea, 97, 214, 219, 225, 228–29
Northern Alliance, Afghan, 186
Northern Command, 6
nuclear exchange calculations, 103–7
nuclear weapons designs, testing issues, 215–26

Office of Management and Budget (OMB), 5, 12
Office of the Secretary of Defense (OSD), 3, 5

Okinawa, 87
On War, Clausewitz's, 1
Operation Desert Storm, 34, 65–67, 71, 75–80, 143–44, 147, 183, 185
orbits, 187–89
ordnance, 183–84
outlays, defense spending, 10–11

Pacific Command (PACCOM), 6, 154–55
Pakistan, 107–10, 114–18
Panama, 34, 82–83
Patriot missile defense system, 201–202, 207–209
Patuxent River Naval Air Station, 3
pensions, military, 39–40
Persian Gulf oil, 51–52
Pierrot, Lane, 65–66
Planning, Programming, Budgeting, and Execution (PPBE) System, 11–12
Pollack, Kenneth, 113
Posen, Barry, 72–80, 117–19, 121
post-traumatic stress disorder (PTSD), 38
program objective memoranda (POMs), 12
propulsion technologies, 180–83
purchasing power parity, 46

Quadrennial Defense Review, 4
Quantico, 3

Radar Horizon, 123, 236
Ramjets, 181
RAND Corporation, 119–20
Reliable replacement warhead (RRW), 223
Revolution in Military Affairs, 171–87, 227, 230–31
robotics, 182–83
rockets, rocket launch costs, 189, 197–98, 202–6
Ronald Reagan Military Buildup, 8, 34, 43
Rumsfeld, Secretary of Defense Donald, 142, 171, 226–27
runway attacks, 132
Russian satellites, 192, 198–99
Rwanda genocide, 151

SL-7 ships, 150
Salvo equation, 92–93
Sandia National Laboratory, 3
satellites, 189
Scott Air Force Base, 3
Second Infantry Division, U.S. Army, 22
Secretary of Defense, 6

selected acquisition reports (SARs), 11, 257–62
sensors, 177–79
shoot-look-shoot tactics in missile defense, 209
simulators, 41
SKEET submunition, 120, 183
Southern Command, 6
space, military uses of, 187–200
space-based laser, 196–97
Special Operations Command, 6
stabilization missions, 107–16
Strategic Command, 6
Strategic Defense Initiative ("Star Wars"), 170–71, 201, 214, 226
Stryker vehicle, 42
submarine-launched ballistic missiles (SLBMs), 103–7, 205–6
submarines, 49–51
Sun Tze, 1, 121, 247
supplemental appropriations, wartime, 12, 16–18

TACWAR, 66
Taiwan-China conflict, 88–103
Taliban, 185
theater high-altitude air defense (THAAD) system, 201–2
Time-Phased Force Deployment List (TPFDL), 142
Transportation Command, 3, 6

V-22 Osprey aircraft, 149
Vandenberg Air Force Base, California, 191
Veterans' Affairs budget, 10
Vick, Alan, 158
Vietnam War, 40, 110–11

Walleye munition, 78
West Point, 37
Wright-Patterson Air Force Base, 3